OXFORD MEDICAL PUBLICATIONS

ASSISTED VENTILATION
AT HOME

ASSISTED VENTILATION AT HOME

A Practical Guide

W. J. M. KINNEAR
University Hospital
Nottingham

Oxford New York Tokyo
OXFORD UNIVERSITY PRESS
1994

Oxford University Press, Walton Street, Oxford OX2 6DP

Oxford New York Toronto
Delhi Bombay Calcutta Madras Karachi
Kuala Lumpur Singapore Hong Kong Tokyo
Nairobi Dar es Salaam Cape Town
Melbourne Auckland Madrid
and associated companies in
Berlin Ibadan

Oxford is a trade mark of Oxford University Press

Published in the United States
by Oxford University Press Inc., New York

A catalogue record for this book is available from the British Library

Library of Congress Cataloging in Publication Data
Kinnear, W. J. M.
Assisted ventilation at home : a practical guide / W.J.M. Kinnear.
p. cm. — (Oxford medical publications)
Includes bibliographical references and index.
1. Respiratory insufficiency—Patients—Home care. 2. Artificial
respiration. I. Title. II. Series.
RC776.R4K56 1994 616.2'004636—dc20 93-33019
ISBN 0 19 262400 8

Typeset by the Electronic Book Factory,
Cowdenbeath, Scotland
Printed and bound in Great Britain by
Biddles Ltd, Guildford and King's Lynn

For Sue and Anne

Preface

Until fairly recently, patients who used ventilators at home were usually cared for by a small number of specialized units. These hospitals have considerable expertise in the field, and will have little need for a book such as this. The physician, anaesthetist, or intensivist who refers a new patient to a specialist unit for initiation of home ventilation may however have had little previous experience, and will probably want to know more about what this involves. If the specialist unit is a long way from the patient's home, their care after discharge will often be shared with local hospital and community services, who need to have a working knowledge of the practicalities of home ventilation. I hope that those who are interested in furthering their knowledge in this area will find this book of value.

More and more hospitals are now setting up their own home assisted ventilation programmes. This is a result of wider dissemination of the necessary expertise and the introduction of less invasive techniques, such as intermittent positive pressure through a nasal mask, but may also reflect the financial advantages of caring for the patient locally. This book is primarily intended as a practical guide for those who are involved in commencing patients on assisted ventilation at home. It is not an exhaustive reference source, and those wishing to know more are referred to the books listed in the bibliography.

Although the text contains a fair amount of medical terminology, much of this will be familiar to patients with chronic respiratory conditions who are about to commence assisted ventilation at home, and I hope that they also will find at least some of the chapters of interest.

W.J.M.K.   Nottingham
June 1993

Acknowledgements

Donald Reinders first encouraged me to write this book, and the project would not have got underway without the enthusiasm of Wendy Reinders. The staff of Oxford University Press subsequently provided valuable guidance. Most of the photographs were taken by the audio-visual department of the University Hospital in Nottingham, and I would like to thank them for their expert help. All the manufacturers listed in Appendix 2 have been generous in lending equipment and providing additional photographs. I would also like to thank Harry Smith for Chapter 8.

Contents

Abbreviations

CO_2	carbon dioxide
CPAP	continuous positive airway pressure
ENPV	external negative pressure ventilation
FEV_1	forced expiratory volume in one second
FiO_2	inspired oxygen concentration
I:E ratio	inspiratory:expiratory ratio
IPPV	intermittent positive pressure ventilation
MEP	maximum expiratory mouth pressure
MIP	maximum inspiratory mouth pressure
n-IPPV	IPPV through a nasal mask
$PaCO_2$	arterial carbon dioxide tension
PaO_2	arterial oxygen tension
PEEP	positive end-expiratory pressure
PEFR	peak expiratory flow rate
t-IPPV	IPPV through a tracheostomy
SaO_2	oxygen saturation
VC	vital capacity

1 Selection of patients for home ventilation

INTRODUCTION

Many different factors must be taken into account when considering a patient for assisted ventilation at home: the disease process which has led to the development of respiratory failure, its likely rate of progression, and the present physiological status of the patient (Table 1.1).

How will assisted ventilation at home help? Randomized clinical trials of assisted ventilation have seldom been performed, but it is usually possible to estimate how assisted ventilation at home will influence the survival of the patient. Even when assisted ventilation is unlikely to prolong life, for instance in progressive neuromuscular diseases, it may be useful in alleviating symptoms such as breathlessness, fatigue, or poor sleep quality.

Assisted ventilation may be indicated on these criteria, but the patient may be mentally or physically unable to cope with the practicalities of using the equipment at home. The level of home support which can be arranged will obviously also be important. Finally, the patient's wishes must be taken into account, after an explanation of what assisted ventilation at home would involve for them and how it would help to solve their problems.

DISEASES LEADING TO VENTILATORY FAILURE

Assisted ventilation at home is used primarily to treat type 2 (hypercapnic) respiratory failure, i.e. failure to maintain adequate alveolar ventilation. The list of causes of ventilatory failure is long, but they can be divided into groups depending on the part of the respiratory system affected. These subdivisions are shown in Fig. 1.1. Not all patients with ventilatory failure will be suitable for assisted ventilation at home, but examples of diseases in each category which do respond well to this treatment are given.

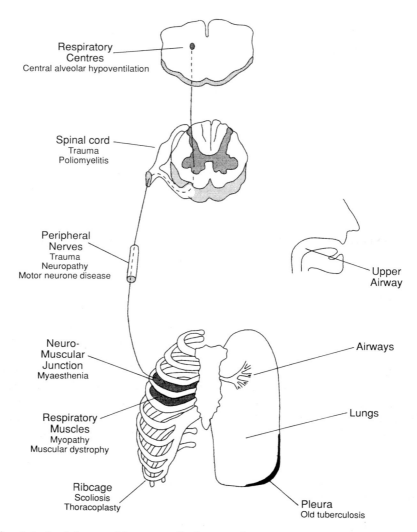

Fig. 1.1 Aetiology of hypoventilation, with examples of diseases suitable for assisted ventilation at home.

Table 1.1 Factors influencing the decision to start assisted ventilation at home

---

Presence of risk factors for developing respiratory failure
Previous episodes of acute respiratory failure
Current respiratory status
Rate of deterioration of respiratory function
Likely effect of assisted ventilation on survival
Likely effect of assisted ventilation on symptoms
Patient's capabilities and home support
Patient's wishes

INDICATIONS FOR ASSISTED VENTILATION AT HOME

Chronic respiratory failure

Table 1.2 shows the main indications for commencing a patient on assisted ventilation at home. Patients who present in stable chronic respiratory failure ($PaCO_2 > 6$ kPa), or with an acute exacerbation of chronic respiratory failure, represent the largest group who will benefit from nocturnal assisted ventilation at home. Evidence from randomized clinical trials is lacking. Nevertheless, the survival of patients with chest wall disease who slide into chronic hypercapnic respiratory failure is generally better in those who receive nocturnal

Table 1.2 Indications for assisted ventilation at home

To prolong life
Chronic hypercapnic respiratory failure
Nocturnal hypoventilation
After recovery from acute respiratory failure: high risk of further life-threatening episodes
Prophylactic: no previous respiratory problems, but high risk

To improve quality of life
Symptoms of diurnal respiratory failure
Cor pulmonale
Sleep disturbance
Ventilator-dependent patient

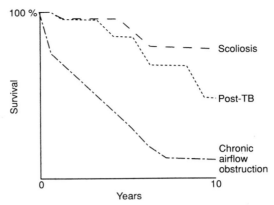

Fig. 1.2 Survival of patients using assisted ventilation at home. (Adapted from *Lyon Medical* 1981;**245**:558, with permission.)

assisted ventilation. Patients with scoliosis, kyphosis, or stable neuromuscular disease do particularly well, but benefit is also seen in patients with a thoracoplasty or other sequelae of tuberculosis (Fig. 1.2). In addition to prolonging life, the patients feel better, have fewer symptoms, and are admitted to hospital with acute exacerbations of their respiratory failure less frequently. The development of cor pulmonale is usually associated with a poor prognosis, and this is an indication to start nocturnal assisted ventilation. Right heart failure usually resolves as chronic hypoxia is corrected, with consequently improved survival. The survival of patients with progressive neuromuscular diseases may not be affected, although their quality of life can be improved (see below). Currently available evidence does not support the use of nocturnal assisted ventilation to improve survival in patients with chronic airflow obstruction.

Nocturnal hypoventilation

Nocturnal hypoventilation is common in patients with chest wall disease, and Fig. 1.3 shows how desaturation during the night can sometimes be predicted from day-time VC. Quite pronounced but transient oxygen desaturation can be seen in patients who are not in day-time respiratory failure and who do not have symptoms related to abnormalities of respiration during sleep (Fig. 1.4). Assisted ventilation is not indicated for these patients, but regular reassessments will be necessary. In chronic respiratory failure, deterioration of gas

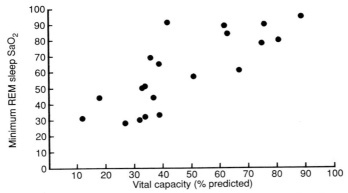

Fig. 1.3 Relationship between VC and nocturnal oxygenation in scoliosis. Desaturation during REM sleep is more pronounced in patients with the lowest VCs (Adapted from *Thorax* 1990;**45**:245 with permission.)

exchange during sleep is common, and usually leads to disruption of the normal sleep pattern. Sleep is therefore unrefreshing, and this contributes to the patient's malaise and tiredness during the day. Nocturnal hypoxia also contributes to their cor pulmonale and polycythaemia.

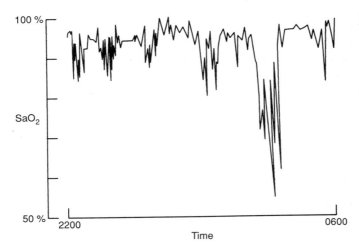

Fig. 1.4 Overnight SaO_2 trace from a patient with muscular dystrophy who did not have day-time respiratory failure or symptoms of sleep disturbance.

Upper airway obstruction should always be excluded as the cause of nocturnal hypoxia. Supplementary oxygen can be administered at night, and if the day-time carbon dioxide tension is normal this is usually well tolerated without inducing hypercapnia. Nevertheless, regular reassessment should be performed, since day-time respiratory failure often develops in subsequent years necessitating nocturnal assisted ventilation.

Acute respiratory failure

Most patients admitted to hospital with acute hypercapnic respiratory failure recover without assisted ventilation, although some will need either non-invasive respiratory support or endotracheal intubation and ventilation in an intensive care unit. Even those patients who require assisted ventilation can usually be weaned over a number of days. On recovery, the circumstances leading to admission should be scrutinized to see if further episodes can be prevented, particularly if the present episode has been life-threatening. Sometimes there is an obvious precipitating factor, such as prescription of night sedation or sedative analgesics, which can be corrected. Airflow obstruction may have been unrecognized or undertreated, particularly if the patient has another respiratory condition to which their breathlessness has been attributed.

After excluding any correctable precipitating factors, an assessment of the risk of further episodes of potentially life-threatening respiratory failure must be made. This will involve consideration of the underlying disease process, its severity and likely rate of progression. It may become apparent that, rather than an isolated episode of acute respiratory failure, this was an acute exacerbation of chronic respiratory failure. If the patient has had more than one episode of hypercapnic respiratory failure within the last year, further episodes are likely. The likely effect of nocturnal assisted ventilation on this risk should be considered. Adequate data are lacking at present, but nocturnal assisted ventilation is probably of most value in patients with scoliosis or stable neuromuscular disease, as is the case when it is used for chronic respiratory failure. In patients with a thoracoplasty, the incidence of further episodes of acute respiratory failure will probably also be reduced, whereas patients with chronic airflow obstruction are unlikely to benefit.

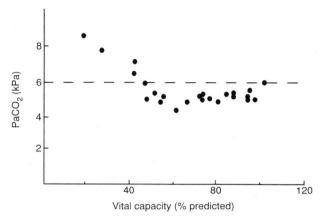

Fig. 1.5 Relationship between VC and $PaCO_2$ in scoliosis. When the VC is below 50% of predicted, $PaCO_2$ starts to rise above the upper limit of normal of 6 kPa. (Adapted from *Thorax* 1991;**46**:476, with permission.)

Prophylaxis

Some patients with minimal respiratory symptoms may nevertheless be at high risk of developing respiratory failure. For example, patients with scoliosis have greatly increased work of breathing, depending upon the severity of their condition. Figure 1.5 shows the relationship between $PaCO_2$ and VC in scoliosis. When VC is less than 50 per cent of predicted values, then the risk of developing hypercapnia is increased. Table 1.3 lists the important factors in assessing the risk of developing respiratory failure in patients with scoliosis.

Although the work of breathing may also be high in patients with neuromuscular problems, hypercapnia is usually only a problem

Table 1.3 Risk factors for the development of acute life-threatening respiratory failure in patients with scoliosis

---

Thoracic curve
Onset of curvature aged <5 years
Curvature >100°
VC <50% predicted or 1 litre
MIP <50% predicted
Associated respiratory muscle weakness (e.g. poliomyelitis)

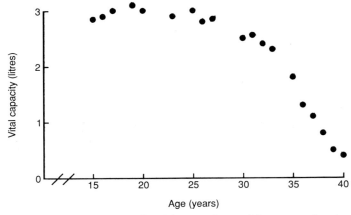

Fig. 1.6 Serial measurements of VC in a patient with a muscular dystrophy, showing an accelerated rate of decline after the age of 30 years. The need for assisted ventilation at the age of 40 years can be predicted several years beforehand.

when the VC is less than 30 per cent of that predicted. The time when this threshold is likely to be crossed can often be predicted from serial measurements, as shown in Fig. 1.6. Indices of respiratory muscle strength can also be used to estimate the risk of respiratory failure. Table 1.4 lists the important factors in assessing the risk of developing respiratory failure in patients with neuromuscular diseases.

Table 1.4 Risk factors for the development of acute life-threatening respiratory failure in patients with neuromuscular diseases

Diaphragmatic weakness
    Clinical evidence
    Standing to supine fall in VC > 15%
    Low transdiaphragmatic pressure
VC <30% predicted or 0.5 litres
Acceleration in rate of decline of VC
MIP <30% predicted
MEP <60 cm H_2O
Poor cough
Associated scoliosis

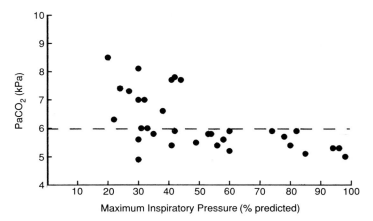

Fig. 1.7 Relationship between MIP and $PaCO_2$ in patients with a thoraco-plasty. There is an increased risk of hypercapnia ($PaCO_2$ >6 kPa) when MIP is less than 50% of predicted.

Respiratory muscle weakness can also be important in other situations. The work of breathing is increased in patients with a thoracoplasty, and Fig. 1.7 shows how these patients are at risk of developing hypercapnia when their maximum inspiratory mouth pressure is less than 50 per cent of that predicted. Airflow obstruction is also important (Table 1.5).

'Prophylactic' nocturnal assisted ventilation has been advocated to try and prevent respiratory failure in patients near or below these thresholds, but there is as yet no evidence that it can achieve this aim. Moreover, in the absence of troublesome symptoms, the

Table 1.5 Risk factors for the development of acute life-threatening respiratory failure in patients with a thoracoplasty

Male sex
Age at operation >30 years
Prior ipsilateral artificial pneumothorax
PEFR < 50% predicted
FEV_1 <1 litre
MIP < 50% predicted

motivation of patients for prophylactic assisted ventilation is usually low. Nevertheless, they should be reassessed regularly and the possibility of assisted ventilation introduced to them. Familiarizing the patient with, for example, n-IPPV or a cuirass at this stage makes the subsequent introduction of assisted ventilation much easier.

Symptomatic relief

Alleviation of day-time symptoms goes hand in hand with the improved survival which is seen with nocturnal assisted ventilation in patients with chest wall disease. As mentioned above, assisted ventilation is worth trying in an attempt to relieve symptoms in patients with progressive neuromuscular diseases whose prognosis is poor. Short periods of assisted ventilation during the day-time may alleviate breathlessness, while nocturnal use will correct sleep fragmentation caused by hypoxia and can also lessen symptoms of cor pulmonale.

The most common group of ventilator-dependent patients are those with a traumatic transection of the spinal cord in the high cervical region. A smaller group of patients with other neurological problems may electively be ventilated as their respiratory function declines. Other patients become ventilator-dependent after an emergency admission to the intensive care unit. When a patient presents to hospital with acute respiratory failure, it is sometimes necessary to intervene with endotracheal intubation and assisted ventilation before all the information necessary to make a decision as to whether this is appropriate or not is available. Patients with poor respiratory function can subsequently be impossible to wean from ventilatory support.

Table 1.6 Factors influencing choice of mode of ventilation

Time ventilatory support needed
Ability to protect airway
Ability to cough
Degree of obesity
Likelihood of upper airway obstruction with ENPV
Thoracic deformity
Age of patient
Nasal problems
Manual dexterity
Preference of patient
Local expertise and resources

Table 1.7 Modes of assisted ventilation

Mode	Indications	Contraindications
t-IPPV	Ventilation >15 hours/day Inability to protect airway Copious respiratory secretions	Poor home support
n-IPPV	Nocturnal ventilation	Nasal blockage Copious secretions Confusion Inability to protect airway Children <5 years Poor manual dexterity Ventilation for >15 hours/day
ENPV	Nocturnal ventilation	Obesity Poor mobility Upper airway obstruction Severe thoracic deformity Ventilation for >15 hours/day
Phrenic pacing	Nocturnal ventilation in central alveolar hypoventilation Daytime ventilation in ventilator-dependent patients	Myopathy Neuropathy Abnormal respiratory mechanics (e.g. scoliosis)
Pneumobelt	Daytime ventilation in ventilator-dependent patients Inability to protect airway	Abnormal respiratory mechanics
Rocking bed	Nocturnal ventilation in diaphragm weakness	Abnormal respiratory mechanics Inability to protect airway

The quality of life for these patients can be dramatically improved by transferring them back to their own homes. Considerable financial savings can also be made and valuable intensive care beds vacated.

CHOOSING THE METHOD OF ASSISTED VENTILATION

The method of assisted ventilation chosen will depend upon a number of factors (Tables 1.6 and 1.7). If ventilatory support is needed for more than 15 hours per day, the patient cannot cough effectively, or they cannot protect their airway, then t-IPPV will be best. Very obese patients may be difficult to fit with a negative pressure ventilator, as may those with severe thoracic deformity. Obesity may also indicate that upper airway obstruction will develop during external negative pressure ventilation, and n-IPPV may be more appropriate. Nasal deformity or obstruction may preclude effective n-IPPV, and some patients with neuromuscular disease lack the manual dexterity to fasten the headgear for a nasal mask, and yet have no problem with the coarser movements needed to fasten a cuirass. In the absence of any contraindications to either technique, patients should be given the opportunity to try external negative pressure ventilation and n-IPPV and make their own choice.

2 *Nasal intermittent positive pressure ventilation*

INTRODUCTION

The technique of inflating the lungs by applying positive pressure to the airway was first introduced towards the end of the last century. Various methods have been used to connect the source of positive pressure, usually a ventilator, to the airway (Table 2.1). Probably the most familiar of these is the endotracheal tube. When this is inserted through the nose or mouth, it can only be left in place for a few days, longer term use requiring a tracheostomy. (A fuller description of long term assisted ventilation through a tracheostomy is given in Chapter 3.)

In conscious patients it is often possible to achieve effective IPPV without intubating the trachea. Face masks and mouthpieces have been used in this context for many years to provide 'non-invasive' assisted ventilation. More recently it was realized that IPPV could be used with a mask, designed for the treatment of obstructive sleep apnoea, which fits over the nose. These masks are comfortable and easy to put on, with the result that n-IPPV has rapidly become the most commonly used form of assisted ventilation for patients at home.

MASKS

Although initially nasal masks were individually constructed for each patient, the range of masks now available commercially is such that it is usually possible to find a comfortable off-the-shelf mask for most

Table 2.1 Methods of administering IPPV

Face mask
Mouthpiece
Endotracheal tube (oral or nasal)
Tracheostomy tube
Nasal mask (n-IPPV)

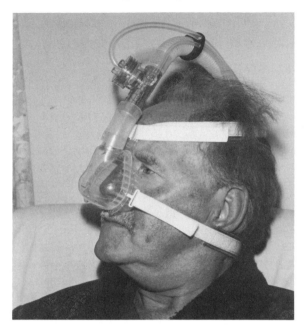

Fig. 2.1 Sefam nasal mask for n-IPPV.

patients. (Details of suppliers can be found in Appendix 2.) No single style of mask matches the needs of all patients. Some prefer to have the tubing running up over their head (Fig. 2.1) rather than down over their chest (Fig. 2.2), and some prefer the firmer foam cushion of the ComfoSeal mask (Fig. 2.3).

Several different designs of headgear are available for holding the mask in place (Figs 2.1–2.3), most of which can be used with types of mask other than that for which they were designed. Most use Velcro fasteners, but the Sefam mask has clips attached to the lower straps which some patients find easier to fasten (Fig. 2.1).

As with the masks themselves, it is probably best to try each type of headgear on each patient and see which they find easiest to use. It is not usually necessary to alter the upper straps on the headgear when the mask is being taken off or put back on again. For those masks which use Velcro for both the upper and lower straps, confusion about which strap fits where on the mask can be avoided by refastening the lower straps in a much looser position before removing the mask over the head.

Patients who cannot tolerate a mask over their face can try n-IPPV

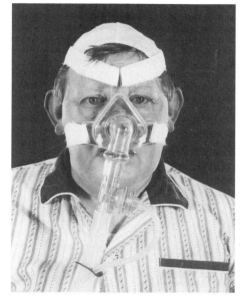

Fig. 2.2 Respironics nasal mask for n-IPPV.

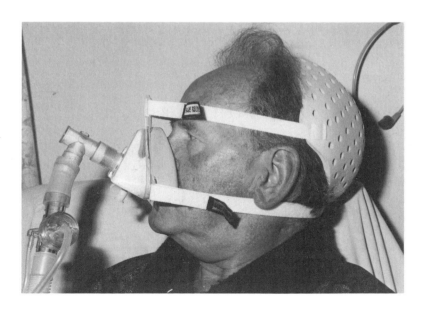

Fig. 2.3 ComfoSeal nasal mask for n-IPPV.

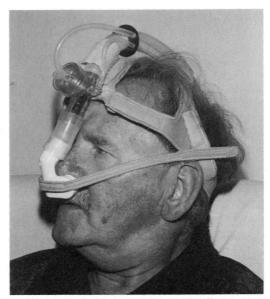

Fig. 2.4 n-IPPV using nasal pillows.

with nasal pillows (Fig. 2.4). This alternative is also a useful temporary measure for patients who develop a pressure sore over the bridge of their nose (see Complications, p.24). The pillows are relatively easily dislodged during the night, however, with consequent loss of efficacy of n-IPPV.

VENTILATORS

Many different ventilators can be set up for n-IPPV, even when they have not been designed primarily for this type of assisted ventilation. Chapter 3 describes ventilators, used for IPPV at home in patients with a tracheostomy, which can easily be adapted for n-IPPV. Two ventilators developed specifically for n-IPPV are described below.

BromptonPac

The BromptonPac ventilator (Fig. 2.5), developed by PneuPac, is a volume cycled flow generator, which can be used with two

Fig. 2.5 The BromptonPac ventilator.

different flow generators. In hospital, a compressed air supply can be attached to the BromptonPac, via an additional unit called an AdapterPac. The AirPac is a separate compressor with which the BromptonPac is used at home, or in hospital if a compressed air supply is not available. Inspiratory and expiratory time are set independently, and the volume delivered is controlled by adjusting the flow rate. The volume delivered to the patient is constant, irrespective of changes in the patient's respiratory compliance and resistance, which can be an advantage when their clinical condition is unstable. During expiration, a pneumatic valve is opened near the connection between the mask and the ventilator tubing, through which the patient exhales (Fig. 2.6). The patient can trigger the BromptonPac, the effort needed to trigger a breath decreasing as the preset expiratory time is approached. A low pressure alarm is accessible at the back of the BromptonPac to detect disconnection or excessive leaks. Table 2.2 and Fig. 2.7 show how to set up the BromptonPac. The BromptonPac measures 22 cm wide by 18 cm high and 23 cm deep, and weighs 4 kg, and the Airpac measures 40 cm wide by 37 cm high and 26 cm deep, and weighs 16 kg.

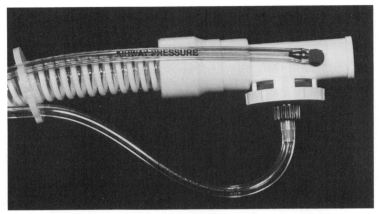

Fig. 2.6 Expiratory valve for BromptonPac circuit.

Nippy

The Nippy, developed by Thomas Respiratory Systems, is a pressure limited ventilator (Fig. 2.8). If a leak develops around the mask or through the patient's mouth, the servo system of the ventilator will increase the flow delivered in order to maintain a constant inflation pressure. Pressure, inspiratory time, and expiratory time

Table 2.2 Setting up the BromptonPac (see Fig. 2.7 for key to controls)

1	Connect the BromptonPac to the AirPac (or AdaptorPac)
2	Attach the ventilator tubing to the BromptonPac
3	Attach mains cable to the AirPac
4	Count patient's breathing rate
5	Calculate time for each breath as follows: Breath time (in seconds) = 60/number of breaths per minute e.g. 15 breaths/minute = 4 seconds breath time, 20/minute = 3 seconds, etc.
6	Set Ti (knob 2) to breath time/2
7	Set Te (knob 3) to breath time/2
8	Set the V_{DEL} (knob 1) to 1.0 litres/second
9	Turn mains on and press On (button 4)
10	Watch the patient's breathing cycle and adjust Ti (knob 2) until the inspiratory time matches the time they take to take a breath in
11	Similarly, adjust Te (knob 3) until the expiratory time matches the time they take to breathe out
12	Increase the Te (knob 3) slightly (by about 0.25 seconds)

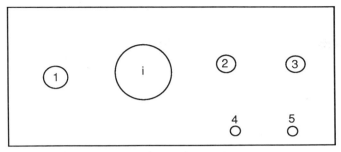

Fig. 2.7 Controls on the BromptonPac.

can be adjusted independently. As with the BromptonPac, the patient exhales through a port in the ventilator tubing which is occluded by a pneumatic valve during inspiration. The patient can trigger the Nippy, and during the first half of expiration the trigger sensitivity is reduced by a factor of three to prevent the ventilator delivering another breath before the patient has finished exhaling. Adjustable alarms can be used to detect low or high pressures. Table 2.3 and Fig. 2.9 show how to set up the Nippy. The Nippy measures 37 cm wide by 26 cm high and 23 cm deep, and weighs 7 kg.

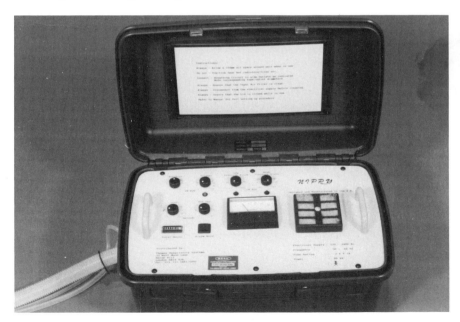

Fig. 2.8 The Nippy ventilator.

Table 2.3 Setting up the Nippy (see Fig. 2.9 for key to controls)

1	Connect ventilator tubing to Nippy
2	Connect mains cable
3	Open lid
4	Set Low Pressure Alarm (knob 3) to 5
5	Set High Pressure Alarm (knob 4) to 30
6	Count patient's breathing rate
7	Calculate time for each breath as follows: Breath time (in seconds) = 60/number of breaths per minute e.g. 15 breaths/minute = 4 seconds breath time, 20/minute = 3 seconds, etc
8	Set Insp (knob 5) to breath time/2
9	Set Max Exp (knob 6) to breath time/2
10	Set Trigger (knob 2) to 0.5
11	Turn power on at side of Nippy
12	Partially cover the end of the tube, or connect to a ventilator bag
13	Adjust the Pressure (knob 1) until the peak (dial i) is 10 cm H_2O
14	Watch the patient's breathing cycle and adjust Insp (knob 5) until the inspiratory time matches the time they take to take a breath in
15	Similarly, adjust Max Exp (knob 6) until the expiratory time matches the time they take to breathe out
16	Increase Max Exp slightly (by about 0.25 of a seconds)

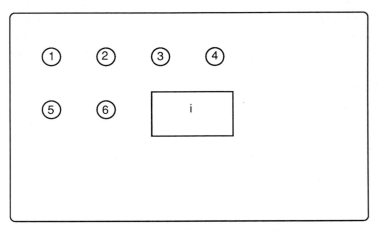

Fig. 2.9 Controls on the Nippy ventilator.

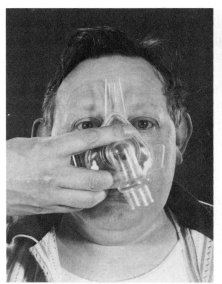 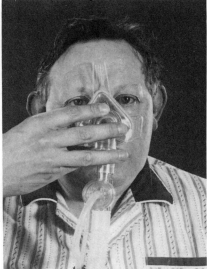

Fig. 2.10 Holding a nasal mask in **Fig. 2.11** Starting n-IPPV. place to accustom the patient.

STARTING A PATIENT ON N-IPPV

After setting up the ventilator the next step in commencing a patient on n-IPPV is to get the patient accustomed to the mask. Select a size of mask which fits fairly tightly but comfortably over their nose, then hold the mask in place gently while the patient breathes quietly through their nose (Fig. 2.10). Repeat this a number of times until they become used to the sensation. Turn the ventilator on and connect it to the mask, then hold the mask in place over the patient's nose for a few breaths (Fig. 2.11). Ask them to keep their mouth closed and breathe with the machine, through their nose. As the patient learns to synchronize with the ventilator, the mask should be held in place for an increasing number of breaths. During this time it may be necessary to gradually increase the amount of air delivered by the ventilator if the patient still has to make a visible effort to breathe. The inspiratory time may also need minor adjustments if the patient feels the breath is too short or too long. If the expiratory time is too short, the patient will not be triggering the ventilator and will probably feel that the next breath is coming too soon. If the expiratory time is too long, the trigger will be less sensitive and a large negative deflection will be seen on the pressure dial at the start of each breath. The expiratory time should then be reduced until this negative deflection is barely noticeable. When the patient can tolerate n-IPPV for several minutes at a time, the mask should be disconnected from the ventilator and strapped in place. The ventilator can then be reconnected to the mask. The low pressure alarm should be set to 5 cm H_2O below the pressure achieved during inspiration, but may need to be lowered if the alarm subsequently goes off frequently despite the patient being ventilated satisfactorily.

The time on n-IPPV can be gradually built up, depending on the urgency of the need for assisted ventilation and the tolerance of the patient. Most patients become accustomed to n-IPPV within a few hours and may be able to fall asleep for an hour or more on the first night they use it. The number of hours can be built up on successive nights, with additional day-time use as necessary. During the first week, the patient should be taught how to attach the mask by themselves, or with the help of a carer if one is to help with this task at home subsequently. Resolution of respiratory failure is often accompanied by a fall in respiratory rate, and small adjustments may therefore be necessary to the inspiratory and expiratory times. The patient may also tolerate a progressively higher tidal volume or inflation pressure if this is necessary to improve gas exchange.

WEANING FROM t-IPPV

n-IPPV can be used to wean patients from conventional IPPV in the intensive care unit. Since sedation is not required during n-IPPV, it is much easier to progressively increase the duration of periods of spontaneous ventilation, compared to a patient who has an endotracheal tube in place. Weaning using n-IPPV is best suited to patients whose sole problem is respiratory failure, and a number of conditions must be satisfied beforehand (Table 2.4). While removal of the endotracheal tube may be associated with a reduction in the volume of respiratory tract secretions, patients who are producing large volumes of sputum are probably better weaned by performing a tracheostomy. Since the airway will be unprotected, it is essential to ensure that the stomach is empty and will remain empty for the first few hours after the endotracheal tube is removed. Adequate function of the gastrointestinal tract must have been demonstrated by absorption of fluid administered via a nasogastric tube (in intubated patients, the absence of bowel sounds does not necessarily mean that the gut is non-functional). No fluid should be put down the nasogastric tube for four hours prior to extubation, and immediately beforehand it should be aspirated and then removed. Care should then be taken to aspirate any secretions in the pharynx. A suction catheter is passed through the endotracheal tube into the trachea and any sputum sucked out, suction on this catheter being maintained as the endotracheal tube is withdrawn. Further pharyngeal aspiration should then be performed. After a few minutes, n-IPPV can be commenced in the same way as described above.

If a tracheostomy has already been performed, n-IPPV can still be used for weaning. Subcutaneous emphysema may occur if the tracheostomy track is not well established, so it is preferable to wait for a few days after performing the tracheostomy before using n-IPPV. A small fenestrated tracheostomy tube should be inserted, the cuff deflated, and the inner cannula removed to minimize resistance to airflow. n-IPPV can then be commenced, and if successful the tracheostomy tube can be removed the following day. After removal of the tube, a firm elastic dressing must be applied to the tracheostomy site to reduce the amount of air leakage.

COMPLICATIONS

n-IPPV can be associated with minor side-effects which patients usually become used to within a few days (Table 2.5). Most patients develop slight reddening of the skin over the bridge of the nose. Care when fastening the upper straps on the headgear is usually sufficient to alleviate this problem, but a spacer on the mask can be used to

Table 2.4 Prerequisites to using n-IPPV to wean patients from t-IPPV

Failed trial of conventional weaning techniques
Ability to sustain spontaneous ventilation for 15 minutes
Intact cough reflex
Minimal sputum production
Functional gastrointestinal tract
Empty stomach
Stable cardiovascular system
Low supplementary oxygen requirements (FiO_2 <0.4)

Table 2.5 Complications of n-IPPV

Nasal pressure sores
Dry nose
Dry eyes
Gastric distension

transfer some of the pressure on to the forehead. A ComfoSeal mask with a notch at its apex may also help this problem. In a few patients it will be necessary to stop n-IPPV for a few days to prevent an ulcer developing, but in most the skin over the bridge of the nose becomes hardened and discomfort soon settles. n-IPPV with nasal pillows (see above) or ENPV (Chapter 4) can be used as a temporary measure if necessary.

A poorly fitting mask or high inflation pressures may lead to air leaking from the edges of the mask which results in dry eyes. Dryness of the nose may also be a problem when the n-IPPV is first started. If high pressures are used, some air may be forced into the stomach and cause uncomfortable distension, but again this usually settles over a few days or weeks.

CONTRAINDICATIONS

The most important disadvantage of n-IPPV is that the airway is not protected and aspiration may therefore occur. For this reason it should not be used in a patient with impaired consciousness (Table 2.6). The only exception to this rule is in patients with moderately abnormal arterial blood gases, in whom the aetiology of their respiratory failure is such that it is likely that they will respond very well to n-IPPV. In such instances, correction of arterial blood gases by n-IPPV, with supplementary oxygen as necessary, may lead

Table 2.6 Contraindications to n-IPPV

Impaired consciousness
Confusion
Type 1 respiratory failure
Respiratory rate >30 breaths/minute
Copious respiratory secretions

to a rapid improvement in conscious level and avoid the need for intubation. It is essential that the patient is monitored closely, and that facilities for intubation are immediately available.

n-IPPV works best in hypercapnic respiratory failure, and is rarely successful in type 1 respiratory failure. Manipulation of the respiratory cycle in an attempt to try and improve oxygenation, for instance by prolonging inspiratory time, usually makes n-IPPV less comfortable and the patient may lose coordination with the ventilator. When the patient's own respiratory rate is over 30 breaths per minute, setting the ventilator to respond to each breath can become difficult. A confused patient will rarely tolerate a nasal mask for long, and is also unlikely to coordinate their respiration with the ventilator. If the patient is producing copious secretions, frequent interruptions of n-IPPV will be necessary for expectoration, and the patient may be better served by intubation. Finally, patients with neuromuscular diseases which involve the facial muscles may develop intolerable air leaks through the mouth. This can be overcome by using a larger mask which fits over the nose and mouth, or changing to ENPV.

MONITORING

The intensity of monitoring during n-IPPV will obviously depend on the context in which it is used. Ideally, transcutaneous carbon dioxide tension should be monitored continuously to ensure that alveolar ventilation is adequate, with oximetry to assess oxygenation. Even then, arterial samples are valuable to monitor changes in acid—base balance, progressive resolution of base excess being a reassuring feature of effective ventilation. In practice, if the patient is well oxygenated during n-IPPV with minimal supplementary oxygen, alveolar ventilation is likely to be adequate and monitoring of carbon dioxide is not essential. (An important exception to this guideline is when supplementary oxygen is being administered, since a small increase in FiO_2 will abolish hypoventilation-induced hypoxia, so that SaO_2 may be normal despite significant alveolar hypoventilation.) Transcutaneous carbon dioxide electrodes are not as widely available as oximeters, but oximetry with periodic arterial blood gas sampling is a satisfactory alternative for most patients.

The levels of oxygen and carbon dioxide during n-IPPV which are deemed satisfactory will depend upon the underlying respiratory condition. Irrespective of the initial level, a progressive fall in $PaCO_2$ indicates that n-IPPV is likely to be effective. (Very occasionally,

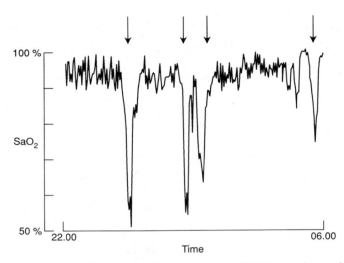

Fig. 2.12 Episodes of desaturation during n-IPPV, resulting from the patient's mouth falling open.

the clinical picture is of improvement, but $PaCO_2$ rises slightly. This is the result of progressive relaxation of the subject, so that they allow the ventilator to do more of the work of breathing and the supplementary action of their own respiratory muscles tails off. Usually the $PaCO_2$ will start to fall shortly thereafter, but obviously careful monitoring of the situation is essential.) Persisting hypoxia and hypercapnia may be improved by increasing the size of each breath. (This is achieved by increasing the flow rate on the BromptonPac or the pressure on the Nippy.) Hypoxia without hypercapnia may necessitate addition of oxygen to the inspired air (see below), but often improves spontaneously as secretions are cleared and atelectasis resolves.

After establishing that n-IPPV is producing adequate gas exchange, the frequency with which reassessments are made will depend on clinical progress. A daily arterial sample will often be sufficient. Non-invasive monitoring during sleep is essential to detect hypoxia or hypercapnia. Periods lasting 20–30 minutes when there is a slight deterioration in gas exchange may be seen, particularly during the latter half of the night. These are related to periods of rapid eye movement sleep, probably associated with a fall in lung volume and change in the pattern of spontaneous respiratory muscle activity. If the clinical condition of the patient is improving, no action is

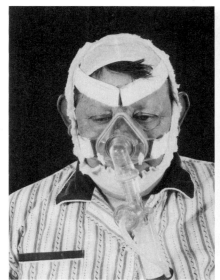
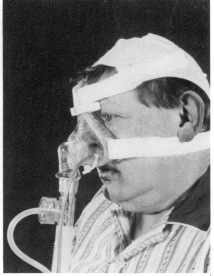

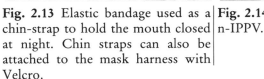

Fig. 2.13 Elastic bandage used as a chin-strap to hold the mouth closed at night. Chin straps can also be attached to the mask harness with Velcro.

Fig. 2.14 Administering oxygen with n-IPPV.

necessary, but a repeat sleep study should be performed a few weeks later. Shorter, more pronounced desaturation may reflect an air leak (Fig. 2.12). If this is seen, direct observation or video should be used to see if this is associated with opening of the mouth, which can be corrected by using a chin strap (Fig. 2.13).

OXYGEN

Administering oxygen to patients with chronic hypercapnic respiratory failure can abolish their hypoxic drive and so further reduce ventilation. When ventilation is being provided by n-IPPV, oxygen can be added safely. It is not necessary to keep to low concentrations of oxygen, and sufficient should be used to correct hypoxaemia. Although oxygen can be added to the inlet of the ventilator, it is generally easier to administer oxygen via a port on the nasal mask (Fig. 2.14) at a flow rate of between 1 and 4 litres per minute depending upon the patient's SaO_2.

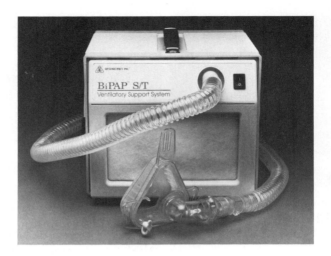

Fig. 2.15 The BiPap ventilator.

BIPAP

Pressure support ventilation is widely used in intensive care units to wean patients from IPPV. BiPap is a version of this technique which can be used to provide assisted ventilation at home through a nasal mask. Two different pressures are set: during inspiration an inflation pressure is used to blow air into the lungs, while during expiration a much lower pressure is used to provide PEEP. This is particularly useful in patients with neuromuscular diseases who only require low inflation pressures, but in whom PEEP is valuable to maintain end-expiratory lung volume and prevent atelectasis. Figure 2.15 shows the Respironics BiPap S/T ventilator. The unit measures

Table 2.7 Setting up the Respironics BiPap unit

Connect tubing to ventilator
Turn mains on
Set mode (knob 1) to Spontaneous/Timed
Set IPAP (knob 2) to 15 cm H_2O
Set EPAP (knob 3) to 2 cm H_2O
Count patient's breathing rate
Set BPM (knob 4) to 2 breaths per minute less than breathing rate

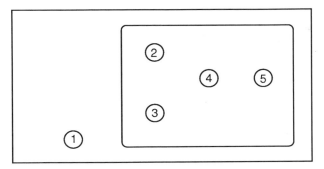

Fig. 2.16 Controls on the BiPap.

20 cm deep by 23 cm high and 31 cm wide and weighs 4 kg. A separate airway pressure monitoring unit is available. Table 2.7 and Fig. 2.16 describe how to set this unit up. After connecting the unit to the patient, the inspiratory pressure should be adjusted until the patient is comfortable and satisfactory ventilation has been established. Expiratory pressure can be increased, within the constraints of what the patient finds comfortable, to improve oxygenation. If satisfactory ventilation cannot be achieved with BiPap, conventional n-IPPV should be tried.

3 Intermittent positive pressure ventilation through a tracheostomy

INTRODUCTION

Tracheostomies have been used for many years to manage patients with obstruction to airflow in the larynx or pharynx. A patient breathing spontaneously through a tracheostomy has a much reduced ventilatory dead space, and this procedure alone can improve gas exchange. It is particularly helpful in patients with a fast respiratory rate, such as those with severe scoliosis or neuromuscular disease, in whom a high proportion of their small tidal volume is normally wasted on dead space ventilation. A tracheostomy also permits easy access to the lower respiratory tract for aspiration of secretions.

During the poliomyelitis epidemics of the 1950s, IPPV through a tracheostomy was developed to treat patients with respiratory muscle paralysis. Many have remained almost constantly on assisted ventilation since then using this technique, which remains the optimum mode of assisted ventilation for patients who are almost totally ventilator dependent.

INDICATIONS

The main indications for t-IPPV are listed in Table 3.1. Most patients who are ventilated using this technique at home first have their tracheostomy stoma formed during an admission to an intensive care unit. An endotracheal tube is initially inserted through the nose or mouth, but the length of admission makes a tracheostomy necessary. Occasionally, it will be clear from the outset that attempts

Table 3.1 Indications for t-IPPV

Assisted ventilation >15 hours/day
Inability to protect airway against aspiration
Ineffective cough
Copious secretions
Inability to provide effective ventilation by less invasive methods

to wean the patient will fail, for example when a progressive decline in lung volumes has been well documented, in a patient with progressive neuromuscular disease, to levels incompatible with spontaneous ventilation. In all other patients, energetic attempts to wean the patient from assisted ventilation must be made. This will usually include periods of mandatory ventilation modalities and pressure support, and even a trial of extubation. ENPV and n-IPPV may also be used as stepping stones to spontaneous ventilation, and even if they fail, they can be tried again after a few weeks or months of t-IPPV.

Other patients may progress from less invasive methods of assisted ventilation to t-IPPV. As neuromuscular diseases progress, weakness of the inspiratory muscles means that the proportion of the day during which assisted ventilation is necessary gradually increases. Once this is more than 12–15 hours, t-IPPV will normally be more convenient. Often expiratory muscle strength also deteriorates. If a tracheostomy has to be performed because the patient can no longer cough effectively, it is generally easier to use this for assisted ventilation also. (Although aspiration of secretions is easy through a tracheostomy, the presence of the tracheostomy tube itself may be the reason that the secretions are produced. Removal of the tube can permit weaning to less invasive methods of assisted ventilation.)

t-IPPV is the most effective form of assisted ventilation, and in some patients, for instance those with severe thoracic deformities, it may be the only method which produces adequate ventilation. Some patients may need to use t-IPPV if they cannot tolerate other methods, perhaps because of nasal symptoms with n-IPPV and difficulty fitting a cuirass.

TRACHEOSTOMY

Tracheostomy tubes

Uncuffed tracheostomy tubes have been used extensively for t-IPPV. They are less likely to lead to local tracheal complications than the early versions of cuffed tubes. The patient can still breathe around the tube if the ventilator fails, but this is of little relevance to modern ventilators which are reliable and fitted with alarm systems. Leakage of air around the tube can be used for speech, but this advantage is offset by dryness of the mouth. This air leakage cannot be quantified, so that careful monitoring of gas exchange is necessary to avoid chronic hypoventilation or hyperventilation. The airway is not protected, so aspiration can easily occur. This may not be important in a patient who is able to protect their airway, for example someone with chronic airflow obstruction who cannot be weaned from a ventilator in the intensive care unit.

t-IPPV is more effectively administered using a modern tracheostomy tube with a low-pressure high-volume cuff. This is much less likely to lead to tracheal damage than older high -pressure cuffs, although it must be appreciated that small volumes of pharyngeal secretions can still be aspirated past such a cuff. If high inflation pressures are used, some air leakage may occur, which can be corrected by increasing the cuff pressure.

Humidification

Since the upper airways are bypassed during t-IPPV, some form of humidification of the inspired air is necessary. Regeneration humidifiers may be used which utilize the heat and moisture contained in the expired air. These are small and convenient to use during the day-time. Cascade humidifiers are bulkier and need a power supply for their heater element. They are more suited to use in the home, particularly at night. The combination of a cascade humidifier at night and a regeneration humidifier during the day will be adequate for most patients. Longer use of the cascade unit may be necessary during lower respiratory tract infections.

Speech

Communication is as important for the ventilator-dependent patient as for anyone else. If the patient is able to breathe spontaneously for a short while and is able to protect their airway, the tracheostomy tube cuff can be deflated, and/or a fenestrated tracheostomy inner tube inserted. (It is important to remember that a suction catheter cannot be passed down a fenestrated inner tube.) In a totally ventilator-dependent patient, effective speech can sometimes be achieved by deflating the cuff and temporarily increasing the amount of air the ventilator delivers to the patient, but there is usually a safer method of voice production. Cuffed tubes can be obtained with an additional port above the cuff through which air can be blown; as this air passes through the larynx it can be used to generate sounds. A sound generator can also be held over the mouth.

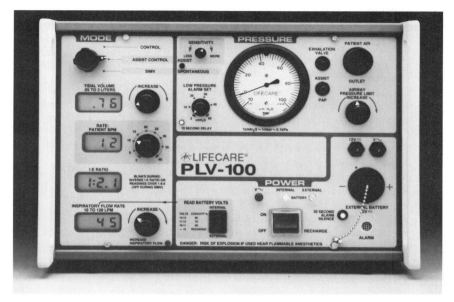

Fig. 3.1 The PLV 100 ventilator.

VENTILATORS

Many different ventilators can be used for t-IPPV, most of which have been designed for use in hospital. Some units specifically designed for home care are described below. Details of suppliers are given in Appendix 2. The manuals supplied with ventilators give details of how to set up each individual machine, but the following section includes summaries of this procedure for the more commonly used ventilators. The assumption is made that the patient is capable of some spontaneous respiration, and that the ventilators will be used in 'assist' mode; some adaptation of the setting-up procedures will be necessary if a fixed rate is to be used.

LifeCare PLV 100

The PLV 100 (Fig. 3.1), like all the ventilators described in this chapter, is volume cycled. It can be powered from the mains supply, with an internal battery backup which lasts for one hour, or from an external battery unit for 24 hours. During inspiration a pneumatic valve, similar to that shown in Fig. 2.6 (p.18), near the connection

Table 3.2 Setting up the PLV 100 (see Fig. 3.2 for key to controls)

1	Turn on at mains and set to Assist Control (knob 1)
2	Connect ventilator tubing to Patient Air port (A)
3	Connect narrower exhalation valve tubing to adjacent Exhalation Valve connector (B)
4	Connect airway pressure tubing to Assist Pap (C) (If the patient is already comfortable on another ventilator, the respiratory rate, I:E ratio and tidal volume can be taken from that ventilator)
5	Count the patient's respiratory rate
6	Set the respiratory Rate (knob 3) on the ventilator to this rate
7	Adjust the Tidal Volume (knob 2) to 0.5 litres
8	Adjust the Inspiratory Flow Rate (knob 4) until the time spent during inspiration and expiration approximately matches the patient's own breathing pattern
9	Adjust the Low Pressure Alarm Set (knob 6) to 5 cm H_2O
10	Partially occlude the ventilator tubing and adjust the Alarm Pressure Limit Increase (knob 8) until the needle on the pressure dial (dial i) does not rise above 40 cm H_2O
11	Set the Sensitivity control (knob 5) to the middle of its range
12	Connect the patient to the ventilator
13	After allowing the patient to settle for a few minutes, increase or decrease the Tidal Volume (knob 2) according to whether they feel that the breath is too deep or too shallow
14	Further adjustments to the Inspiratory Flow Rate (knob 4) can be made until the inspiratory pattern feels most natural for the patient
15	Decrease the Respiratory Rate (knob 3) by 2–3 b.p.m.
16	Turn the Sensitivity (knob 5) down until the patient can trigger breaths comfortably
17	Set the Low Pressure Alarm Set (knob 6) to 5 cm H_2O below the peak inflation pressure (dial i)
18	Disconnect the patient from the ventilator
19	Partially occlude the ventilator tubing and set Airway Pressure Limit Increase (knob 8) so that the pressure (dial i) does not exceed 10 cm H_2O above the peak inflation pressure noted previously
20	Reconnect to patient

of the ventilator tubing to the tracheostomy, is occluded and air is pushed into the patient's lungs. During expiration, the pneumatic valve is released and the patient exhales through it.

Tidal volume, respiratory rate, and I:E ratio can be adjusted independently, as can the inspiratory flow rate within the constraints of the inspiratory time and tidal volume. In the assist/control mode the patient can trigger breaths. There is an adjustable limit on maximum airway pressure and a low pressure alarm detects disconnection.

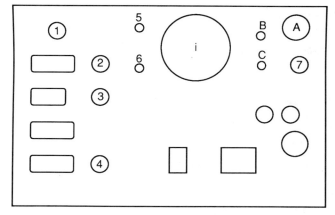

Fig. 3.2 Controls on the PLV 100.

There is also a remote alarm facility. The PLV 100 weighs 13 kg and measures 23 cm high by 31 cm wide and 31 cm deep.

Table 3.2 and Fig. 3.2 describe how to set up the PLV 100.

Companion 2801

The Puritan Bennett Companion 2801 (Fig. 3.3) is similar to the PLV 100 in that it is a volume cycled ventilator which can be powered from AC mains, from an internal battery for up to one hour, or from an external DC battery. A pneumatic valve on the ventilator tubing controls expiration and inspiration, similar to that described above. A gas collector head can be connected to the pneumatic valve, for analysis of expired air. There is a sigh facility, a slide rail for attaching a cascade humidifier, and a remote alarm connector. The Companion 2801 weighs 16 kg and measures 27 cm high by 32 cm wide and 34 cm deep.

Table 3.3 and Fig. 3.4 describe how to set up the 2801.

Monnal D

The Monnal D (Fig. 3.5) is an unsophisticated volume-cycled ventilator with a long track record of safety and reliability for home ventilation. It works well for patients with neuromuscular diseases, but can lack the power to ventilate patients with chest wall deformity effectively. A pneumatic valve on the ventilator tubing controls expiration and inspiration, similar to that described above. It can only be

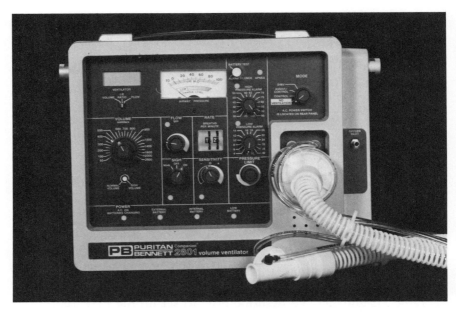

Fig. 3.3 The Companion 2801 ventilator.

powered from AC mains and there is no remote alarm facility. Stacking of breaths can occur since there is no protected expiratory time during which another breath cannot be triggered. The Monnal D weighs 14 kg and measures 16 cm high by 47 cm wide by 31 cm deep.

Table 3.4 and Fig. 3.6 describe how to set up the Monnal D.

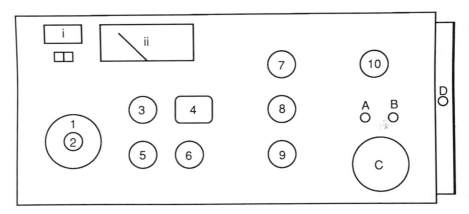

Fig. 3.4 Controls on the 2801.

Table 3.3 Setting up the Companion 2801 (see Fig. 3.4 for key to controls)

1	Connect ventilator tubing to Ventilator Outlet (C)
2	Connect exhalation valve to Exhalation Valve connector (B)
3	Connect airway pressure connector to Airway Pressure (A)
4	Turn power on and set Mode (knob 10) to Assist/Control
5	Set Sigh (knob 5) to off
6	Turn Pressure Limit (knob 9) fully clockwise
7	Turn High Pressure Alarm (knob 7) to 100 cm H_2O
8	Turn Low Pressure Alarm (knob 8) to 2 cm H_2O (If the patient is already comfortable on another ventilator, the respiratory rate, I:E ratio, and tidal volume can be taken from that ventilator)
9	Count the patient's respiratory rate
10	Set the Respiratory Rate (switch 4) to this rate
11	Set Flow (knob 3) to the midpoint (12 o'clock on scale)
12	Set Normal Volume and Sigh Volume (knobs 2a and b) to 500 ml
13	Adjust the Flow (knob 3) on the ventilator until the time spent during inspiration and expiration approximately matches the patient's own breathing pattern
14	Connect the patient to the ventilator
15	After allowing the patient to settle for a few minutes, increase or decrease the Volume (knob 2a) according to whether they feel that the breath is too deep or too shallow
16	Adjust the Flow (knob 3) until the inspiratory pattern feels most natural for the patients
17	Decrease the Rate (switch 4) by 2–3 b.p.m.
18	Adjust Sensitivity (knob 6) until the patient can trigger breaths comfortably (usually just to the − side of '0')
19	Set the Low Pressure Alarm (knob 8) to 5–10 cm H_2O below the peak inflation pressure (display ii)
20	Set the High Pressure Alarm (knob 7) to 10 cm H_2O above the peak inflation pressure
21	Disconnect the patient from the ventilator
22	Partially occlude the ventilator tubing and turn Airway Pressure Limit (knob 9) anticlockwise until the pressure (display ii) does not exceed 10 cm H_2O above the peak inflation pressure noted previously
23	Reconnect to patient

Monnal DCC

The Monnal DCC (Fig. 3.7) is a more modern version of the Monnal D. Unlike those of the other ventilators described in this chapter, the expiratory valve is incorporated in an expiratory block which is attached to the ventilator. This has the advantage that expiratory

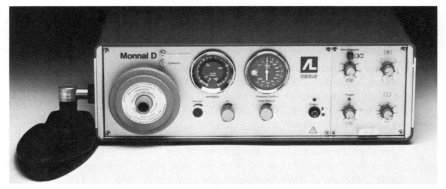

Fig. 3.5 The Monnal D ventilator.

airflow is monitored routinely, but this does mean that an expiratory and inspiratory tube are necessary. PEEP can easily be added through the expiratory valve. An optional oxygen analyser can be fitted, making monitoring of oxygen therapy easy. There is an internal battery, with the facility for connecting an external battery. Sighs can be incorporated and there is a remote alarm facility. Ventilatory parameters can be monitored externally, accessible through an RS232 compatible interface. The DCC weighs 22 Kg and measures 22 cm high by 39 cm wide and 41 cm deep.

Table 3.5 and Fig. 3.8 describe how to set up the Monnal DCC.

CHOOSING A VENTILATOR

Several factors must be taken into account when selecting a ventilator for an individual patient to use at home (Table 3.6). If the patient

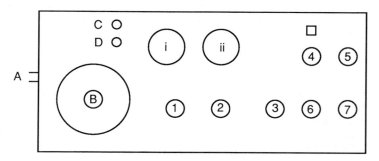

Fig. 3.6 Controls on the Monnal D.

cannot breathe for more than a few minutes without assisted ventilation, it is essential to have a ventilator with an internal battery which can be used in the event of an interruption to the mains power supply. The PLV 100, Monnal DCC, and Companion 2801 all have this facility.

If assisted ventilation is needed during the day-time, mobility will be increased by the use of a ventilator with an external battery connection which can be fitted to a wheelchair. (On motorized wheelchairs, it is important that the battery used for the ventilator is not the same one that powers the wheelchair.) The PLV 100, Monnal DCC, and Companion 2801 all have external battery connections. The PLV 100 and Companion 2801 are smaller and lighter than the Monnal DCC.

Most patients using t-IPPV can be ventilated on air. If supplementary oxygen is necessary (see below), monitoring of FiO_2 is easiest with the Monnal DCC. The PLV 102 has an oxygen conserving

Table 3.4 Setting up the Monnal D (see Fig. 3.6 for key to controls)

1	Connect bag to side of ventilator (A)
2	Connect ventilator tubing to ventilator output (B)
3	Connect airway pressure tubing to Airway Pressure (C)
4	Connect expiration valve to Collector (D)
5	Turn power on (knob 3)
6	Turn Air (knob 1) to 15 litres/minute
7	Set Minipressure (knob 4) to 5 (If the patient is already comfortable on another ventilator, the respiratory rate, I:E ratio, and tidal volume can be taken from that ventilator)
8	Establish the patient's respiratory rate by counting the number of breaths taken over one minute
9	Set frequency (knob 7) to the patient's rate
10	Still without connecting the patient to the ventilator, adjust the I:E Ratio (knob 5) on the ventilator until the time spent during inspiration and expiration approximately matches the patient's I:E ratio
11	Set trigger (knob 6) to 0
12	Connect to patient
13	Adjust Air (knob 1) until tidal volume feels comfortable for the patient
14	Watch the bag: increase Air (knob 1) if bag empties completely during inspiration, decrease Air if bag becomes over -distended during expiration
15	Observe maximum inflation pressure (dial ii) achieved during inspiration
16	Disconnect patient from ventilator
17	Partially occlude the ventilator tubing and set Safety Pressure (knob 2) to 10 cm H_2O above the maximum inflation pressure
18	Reconnect to patient

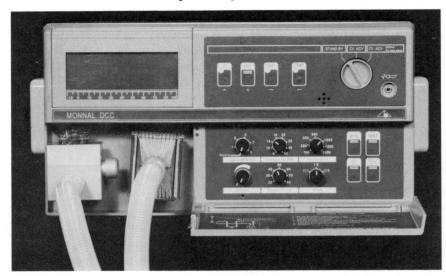

Fig. 3.7 The Monnal DCC ventilator.

facility which may make an oxygen cylinder last longer. Although the Monnal DCC has the disadvantage of inspiratory and expiratory tubes connected to the patient, this allows expiration to be controlled more easily and some patients find this ventilator more comfortable.

If none of the above factors are critical to the choice of ventilator, the familiarity of staff with the equipment may be important. In any unit, use of one particular make of ventilator will improve the

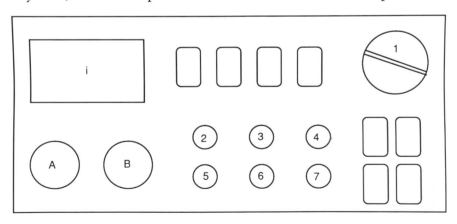

Fig. 3.8 Controls on the Monnal DCC.

Table 3.5 Setting up the Monnal DCC (see Fig. 3.8 for key to controls)

1	Set mode (knob 1) to assisted controlled ventilation (ACV)
2	Establish the patient's respiratory rate by counting the number of breaths taken over one minute (If the patient is already comfortable on another ventilator, the respiratory rate, I:E ratio, and tidal volume can be taken from that ventilator
3	Set Frequency (knob 3) to this rate
4	Set Tidal Volume (knob 4) to 500 ml
5	Still without connecting the patient to the ventilator, adjust the Flow (knob 7) on the ventilator until the time spent during inspiration and expiration approximately matches the patient's own breathing pattern
6	Set Max Pressure (knob 6) to 50 cm H_2O
7	Set Peep (knob 5) to 0
8	Connect the patient to the ventilator
9	After allowing the patient to settle for a few minutes, increase or decrease the Volume (knob 4) according to whether they feel that the breath is too deep or too shallow
10	Adjust I:E Ratio (knob 7) until the inspiratory pattern feels most natural for the patient
11	Decrease the Rate (switch 4) by 2–3 b.p.m.
12	Adjust Sensitivity (knob 6) until the patient can trigger breaths comfortably

confidence of staff in using the ventilators. If a patient develops problems at home, a single backup ventilator can be substituted.

STARTING A PATIENT ON t-IPPV

For most patients, t-IPPV is commenced on the intensive care unit after per-oral or per-nasal endotracheal intubation for an episode of acute respiratory failure. A tracheostomy will usually be performed after 7–10 days, and thereafter the transition to t-IPPV is relatively easy. Sedation is no longer needed and the amount of time spent off the ventilator can be gradually increased to establish the individual requirements of each patient. In some, assisted ventilation may only be necessary at night, while others may need additional periods of t-IPPV during the day. If periods of spontaneous ventilation can be established, even if only for a few minutes each hour, this makes subsequent care of the patient at home easier.

The transition to t-IPPV can be very traumatic for a patient with chest wall disease in whom n-IPPV and/or ENPV have failed.

Table 3.6 Choosing a ventilator

Is the patient ventilator-dependent?
Is t-IPPV needed during the day-time?
Is supplementary oxygen required?
Is PEEP necessary?
Is a remote alarm necessary?
With which ventilators are staff familiar?
What back-up is provided?
What is the cost?

The decision to proceed to a tracheostomy can usually be made semi-electively, in consultation with the patient and their family and carers. Fear of loss of independence and dependence on the ventilator are common, and may be alleviated to some extent by careful explanation beforehand. It can be helpful for the patient to meet someone else who already uses t-IPPV at home. They should be warned about the initial discomfort of the tracheostomy post-operatively, and about sucking-out procedures. Plans should also be made for speech production and explained to the patient.

COMPLICATIONS

Local complications of the tracheostomy are rare with modern tubes, but tracheal stenosis, ulceration, and haemorrhage can still occur. A tracheostomy gives bacteria access to the lower respiratory tract, so infections are more common. In contrast to n-IPPV, t-IPPV is a sealed system, so excessive airway pressures can be generated which can lead to barotrauma and pneumothorax (Table 3.7).

Table 3.7 Complications of t-IPPV

Tracheal ulceration
Tracheal haemorrhage
Tracheal stenosis
Lower respiratory tract infections
Barotrauma
Pneumothorax

CONTRAINDICATIONS

Tracheostomy is contraindicated in patients with a bleeding diathesis, or a local problem such as a goitre. t-IPPV is more likely to produce pneumothoraces in patients with bullous lung disease, in whom inflation pressures should be kept as low as possible.

MONITORING

As with other forms of assisted ventilation, the frequency with which gas exchange is monitored will depend upon the clinical situation. Effective ventilation can usually be established easily with t-IPPV, and care must be taken not to hyperventilate the patient. If arterial blood gas sampling shows adequate alveolar ventilation and oxygenation, overnight monitoring is usually unnecessary, since leaks are unlikely to develop during the night, as is the case with n-IPPV or a cuirass.

SUPPLEMENTARY OXYGEN

If supplementary oxygen is needed to maintain a satisfactory SaO_2, the Monnal D, Monnal DCC and Companion 2801 have inlet ports where oxygen can be added. Tables are provided to calculate the rate of oxygen administration to achieve the FiO_2 required, but this should always be checked with an oxygen analyser at the patient outlet. The Monnal DCC can be supplied with an internal oxygen analyser, which makes precise adjustment of FiO_2 easy.

Oxygen can be bled into the patient circuit used with the PLV 100, but this will increase tidal volume. The PLV 102 is similar to the PLV 100, but it is fitted with an oxygen inlet. FiO_2 can be adjusted using a knob on the front of the ventilator, but an external oxygen analyser should be used to check this. The PLV 102 also includes a sigh facility which is not present on the PLV 100.

PEEP

If satisfactory oxygenation during t-IPPV is difficult to achieve, PEEP can be applied. This can easily be adjusted on the Monnal DCC, but it can be achieved on the other ventilators by fitting an external valve to the expiration ports.

4 External negative pressure ventilation

BACKGROUND

During the last century, several attempts were made to assist respiration by enclosing the body, or sometimes just the chest, up to the neck in a rigid structure. Creating a vacuum inside this structure produced a pressure gradient between the thorax and the mouth, which resulted in airflow into the lungs. Further development work took place in the first half of this century, usually driven by the stimulus of poliomyelitis epidemics. The more successful types of apparatus for treating patients with respiratory muscle paralysis caused by acute poliomyelitis were various designs of cabinet respirators, or iron lungs. As mentioned in Chapter 3, during the 1950s IPPV through a tracheostomy became the method of choice for artificial ventilation, but cabinet respirators have continued to be used to manage a small number of patients. Figure 4.1 shows a modern version of the cabinet respirator. Smaller cuirass and jacket respirators were valuable for patients with poliomyelitis and less severe ventilatory problems, or during convalescence. Although less effective than cabinet respirators, these are more convenient for the patient to use at home and have proved useful for nocturnal assisted ventilation in patients with chest wall deformities or neuromuscular diseases.

PATIENTS

External negative pressure ventilation, like n-IPPV, does not require sedation of the patient. It therefore has similar advantages over endotracheal intubation for the treatment of acute respiratory failure, as discussed in Chapter 2. The cabinet respirator is the best external pressure device for use in this situation (see below). As with n-IPPV, there is a danger of aspiration in patients with impaired consciousness or neuromuscular problems who are not able to protect their airways. The degree of cooperation required of the patient can be less than that needed for n-IPPV, so patients with acute respiratory failure who cannot tolerate a nasal mask sometimes do very well in a cabinet respirator. This may be related to the fact that incoordination with the respirator is less disruptive, and speech is much easier.

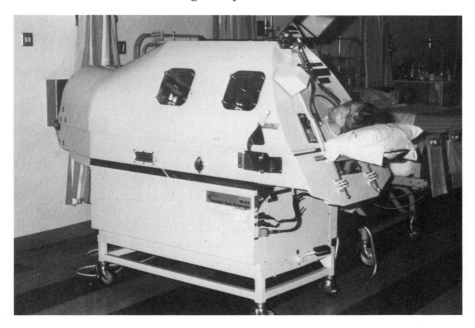

Fig. 4.1 A modern cabinet respirator.

Some patients with a severe thoracic kyphosis do not have sufficient room between their head and shoulders to accommodate the neck seal of a cabinet respirator, and the presence of a tracheostomy makes the use of this technique more difficult but not impossible. Very obese patients may be difficult to get in and out of the cabinet, and these patients are particularly prone to upper airway obstruction during external negative pressure ventilation (see below).

Cabinet respirators are cumbersome for use at home, and smaller jacket or cuirass respirators are more commonly used in this context. Patients differ in whether they object more to a nasal mask or to the restriction of a jacket or cuirass. External negative pressure devices can be valuable when nasal pressure sores or nasal dryness arise during n-IPPV, and are generally tolerated better than a nasal mask in children under the age of five years. Both jacket and cuirass respirators require some mobility and strength on the part of the patient if they are to be fitted at home without assistance, but the movements require less dexterity than fitting a nasal mask. Depending upon the design of the seal, a jacket respirator can be impractical for a patient with a tracheostomy, and severe thoracic deformity can make manufacture of an effective cuirass difficult.

PUMPS

External negative pressure ventilation is associated with leaks, the flow rate through which, although variable between different designs of apparatus, is generally high. Moreover, the leakage flow rate can vary considerably over short periods of time, for instance if the patient changes position. The ideal pump for external negative pressure ventilation is a pressure cycled machine which is designed to maintain the same peak pressure within the external negative pressure device in the face of variable leaks.

Many bellows pumps with a fixed stroke volume are still in use with different external negative pressure devices. These pumps are rugged and reliable, but they cannot compensate for changes in the leakage flow rate and they are relatively inflexible. Pumps currently manufactured for ENPV are described below.

Fig. 4.2 The Newmarket negative pressure ventilator.

The Newmarket pump

This pump (Fig. 4.2) is a flow generator, with a maximum flow capability of approximately 1000 litres/minute. This flow capability allows it to be used with all types of external negative pressure respirator. It is servo-controlled to maintain a constant peak external negative pressure. Respiratory rate, I:E ratio, and pressure can all be controlled independently. An optional facility enables an external signal to be used to control the servo mechanism of the pump, which can be used to drive the pump from an external signal generator, such as a microcomputer or a triggering device. Separate units can be purchased for triggering from an oro-nasal pressure sensor, or to superimpose continuous external negative pressure on the cyclical pressure waveform (analagous to PEEP during IPPV). The external input can also be used for electronic synchronization of two Newmarket pumps, as may be needed for particularly large cabinet respirators with high leakage flow rates. Alarms are fitted to detect loss of mains power or a drop in pressure as a result of disconnection of the tubing or the development of excessive leaks for which the pump cannot compensate. The pump measures 88 cm high by 33 cm wide and 38 cm deep and weighs 34 kg. Table 4.1 and Fig. 4.3 outline the steps in setting up the Newmarket pump for ENPV. Figure 4.4 shows the smaller, compact version of the Newmarket pump, for use with a cuirass or jacket respirator. This measures 16 cm high by 33 cm deep and 43 cm wide and weighs 14 kg. Setting up this pump is the same as for the Newmarket pump.

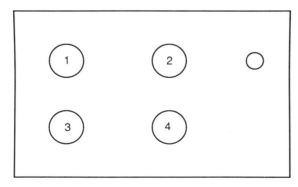

Fig. 4.3 Controls on the Newmarket ventilator.

Table 4.1 Setting up the Newmarket pump (see Fig. 4.3 for key to controls)

1	Connect hose to pump inlet
2	Turn mains on (switch on top of pump)
3	Set positive pressure (knob 4) to 0
4	Set negative pressure (knob 3) to 20
5	Count the patient's respiratory rate
6	Set rate (knob 1) to this rate
7	Adjust the I:E ratio (knob 2) until the inspiratory/ expiratory ratio of the pump matches that of the patient
8	Connect hose to cuirass/jacket/cabinet
9	Adjust negative pressure (knob 3) until the patient feels that they are getting a deep enough breath from the ventilator. As they settle, any visible accessory muscle contraction during inspiration should disappear
10	Adjust rate until the patient feels comfortable, often at a slightly lower rate than their spontaneous rate
11	Adjust the I:E ratio if necessary
12	Monitor gas exchange and increase negative pressure (knob 3) if inadequate alveolar ventilation

The Emerson negative pressure chest respirator

This ventilator (Fig. 4.5) has fixed rate (control) and assist modes, the latter utilizing pressure sensed from nasal cannulae as the trigger signal. Breaths can also be triggered manually. Respiratory rate and inspiratory time can be set independently, allowing manipulation of the I:E ratio. Peak external negative pressure can be controlled, and continuous external negative pressure superimposed. A monitoring tube accompanies the main ventilator tube, giving more accurate setting of the pressure inside the negative pressure device. The pump measures 31 cm high by 41 cm wide and 28 cm deep and weighs 12 kg. Table 4.2 and Fig. 4.6 describe how to set up the Emerson pump.

NEV 100

LifeCare have recently introduced a servo-controlled negative pressure ventilator, the NEV 100 (Fig. 4.7). This has a display screen with menus to set the microprocessor which controls the pump and the alarm systems. Triggering, sighs, and continuous external negative pressure can all easily be set from the menus. The display also gives digital and graphical displays of the ventilation parameters during ENPV. The NEV 100 measures 30 cm high by 30 cm wide and 50

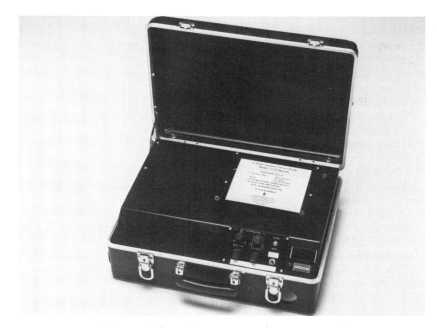

Fig. 4.4 The SiPlan compact cuirass ventilator.

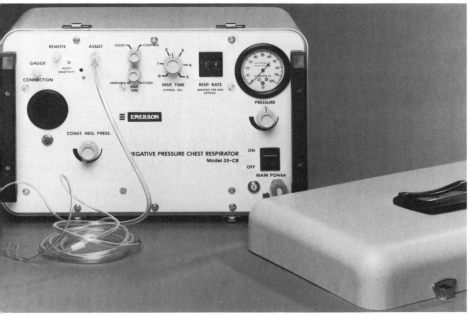

Fig. 4.5 The Emerson negative pressure ventilator.

Table 4.2 Setting up the Emerson pump (see Fig. 4.6 for key to controls)

1	Connect the hose to the pump (inlets A and B) with the hose disconnected from the patient's ENPV device
2	Turn power on (switch 7)
3	Set mode (knob 1) to 'Control'
4	Set Pressure (knob 6) to mid-point of scale
5	Turn Cont Neg Press (knob 5) fully anticlockwise
6	Set Insp Time (knob 2) to 'Fixed'
7	Count the patient's respiratory rate
8	Set the Resp Rate (switch 4) to this rate
9	Adjust Insp Time (knob 3) until the proportion of each breath spent in inspiration matches the patient's own breathing pattern
10	Connect hose to cuirass/jacket/cabinet
11	Adjust Pressure (knob 6) until the patient feels that they are getting a deep enough breath from the ventilator. As they settle, any visible accessory muscle contraction during inspiration should disappear
12	Adjust Resp Rate (switch 4) until the patient feels comfortable, often at a slightly lower rate than their own spontaneous rate
13	If necessary, adjust the Insp Time (knob 3) until the inspiratory—expiratory ratio feels most comfortable for the patient
14	Monitor gas exchange and increase Pressure (knob 6) if inadequate alveolar ventilation

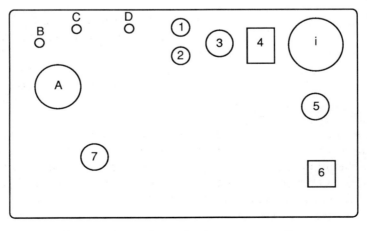

Fig. 4.6 Controls on the Emerson ventilator.

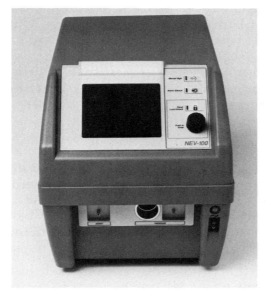

Fig. 4.7 The NEV 100 negative pressure ventilator.

cm deep and weighs 14 kg. The procedure for setting is menu-driven after switching on the ventilator (Table 4.3).

RESPIRATORS

Cabinet respirators

The most effective form of ENPV respirator is a rigid chamber which encloses all the body from the neck down. Older cabinet (tank) respirators were constructed from wood or metal. Many of these rugged iron lungs remain in service decades after their construction. More modern versions are usually constructed from fibreglass or Perspex, and hence are lighter and more portable. An example is shown in Fig. 4.8.

Patients may initially be intimidated by the appearance of a cabinet respirator, or may become more breathless lying flat in the cabinet before the pump is started. They need to be reassured and encouraged as ENPV is commenced, and most patients settle very rapidly.

Obtaining a good seal around the patient's neck is the most difficult part of using a cabinet respirator. Some cabinets have a fixed end, with a diaphragm through which the patient inserts their head. This diaphragm can be tightened to produce a seal. Other cabinets are made with a lid which hinges open. After the two halves are closed, soft pieces of rubber are positioned around the neck and clamped on to the cabinet. With either of these styles of neck seal, additional padding may be necessary to reduce air leaks, and patients sometimes find it more comfortable to put on a soft cervical collar beforehand.

Jacket respirators

During ENPV, movement of only the ribcage and abdomen contributes to respiration. Creating a negative pressure around the rest of the body, as in a cabinet respirator, is inefficient and may occasionally be harmful if it leads to peripheral pooling of blood. In a jacket respirator, the limbs are not included within the rigid portion of the respirator (Fig. 4.9). This means that the respirator is smaller,

Table 4.3 Setting up the NEV 100

1	Connect ventilator tubing and pressure monitoring tubing to front of ventilator
2	Switch ventilator on
3	Press Panel Lock/Unlock key once to access menus
4	Using control knob, set Mode to 'Control'
5	Count the patient's spontaneous respiratory rate, and set Rate to this rate
6	Set Negative Pressure to -30 cmH$_2$O
7	Set Base Pressure to 0 cmH$_2$O
8	Adjust I:E Ratio until the pattern of the ventilator corresponds to the patient's spontaneous breathing pattern
9	Set Low Pressure alarm to 0 cmH$_2$O
10	Connect the ventilator to the patient in their cuirass/jacket/cabinet
11	Adjust Negative Pressure until the patient feels that they are getting a deep enough breath from the ventilator. As they settle, any visible accessory muscle contraction during inspiration should disappear (If more than 30 seconds between adjustments in steps 11–15, press Panel Lock/Unlock key once to access menus again)
12	Adjust Rate until the patient feels comfortable, often at a slightly lower rate than their own spontaneous rate
13	If necessary, adjust the I:E Ratio until the breathing pattern feels most comfortable for the patient
14	Monitor gas exchange and increase Negative Pressure if inadequate alveolar ventilation
15	Set Low Pressure alarm to 10 cmH$_2$O above Negative Pressure (eg -20 cmH$_2$O if Negative Pressure is -30 cmH$_2$O)

Fig. 4.8 The PortaLung cabinet respirator.

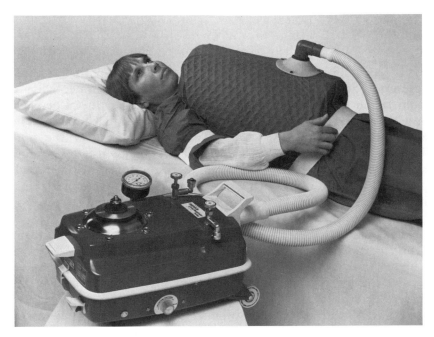

Fig. 4.9 The Emerson jacket respirator.

Table 4.4 Making a jacket respirator

1	Make two simple Stockinette vests and put on the patient
2	Position the patient on a couch, lying as flat as they can
3	Using wide strips of plaster, make a cast of the patient's thorax, including as much as possible of the chest and abdomen from the sternal angle to the iliac crests and extending down on to the sides of the bed laterally (Fig. 4.10)
4	Remove the cast from the patient, cutting the Stockinette vests, and build it up by 5–10 cm all over, with the greatest thickness over the part which will cover the lower ribcage and abdomen (Fig. 4.11). Foam rubber or other material can be used to build up the cast, with a top layer of plaster
5	Use X-lite themolabile plastic to make the rigid shell from the plaster cast. Four or five layers of X-lite will be needed to give the necessary rigidity
6	Remove the mesh frame from the cast and cover the edges with foam (Fig. 4.12)
7	Cut a hole in the centre of the front of the frame and insert a plastic plumbing connector (Fig. 4.13)
8	Choose an airtight jacket large enough to cover the shell
9	Stick Velcro to the upper arms to make an airtight fastening
10	Cut a hole in the centre of the jacket, slightly larger than the connector
11	Cut a piece of Neoprene, with a hole in the centre slightly smaller than the connector. Glue this to the jacket, over the hole made in stage 10, with contact adhesive
12	Fit self-retaining fasteners to the neck and waist cords

easily portable, and requires a less powerful pump than a cabinet respirator. Several jacket respirators are available commercially. The rigid frame which fits over the chest can be either metal or plastic, and may be used as a single piece or with a back plate. There are also several styles of airtight garment to fit over the frame. Whereas in a cabinet the most difficult problem is usually obtaining an adequate neck seal, with a jacket respirator the lower seal is generally the more problematic. Some jackets overcome this by including the legs in the garment, which then resembles a sleeping bag or wetsuit.

Jacket respirators are easy to make, using equipment which can be found in most occupational therapy or physiotherapy departments. The materials used are inexpensive, and the respirator can be tailored to suit the individual patient. The steps in construction of a jacket respirator are listed in Table 4.4 and shown in Fig. 4.10–4.14. To start ENPV with a jacket respirator, put the jacket on the patient, and insert the frame inside the jacket. Tighten the arm bands, neck

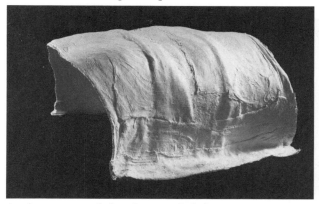

Fig. 4.10 A plaster cast of the chest and abdomen to act as a mould for a cuirass or jacket respirator shell.

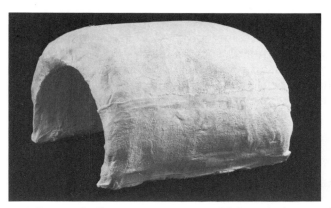

Fig. 4.11 The cast built up to act as a mould for a jacket respirator.

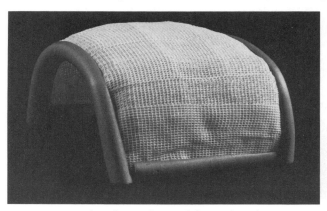

Fig. 4.12 X-lite frame formed from the plaster cast.

Fig. 4.13 Plumbing connector for negative pressure ventilator tube.

seal, and waist band, connect the pump, and commence ENPV. Normally a peak pressure of 15–30 cm H$_2$O below atmospheric pressure will be required. If excessive airleaks round the neck or waist are uncomfortable or prevent an adequate pressure being obtained, a foam collar or belt can be attached to the patient before the jacket is put on. If the frame touches the patient during ENPV, remove

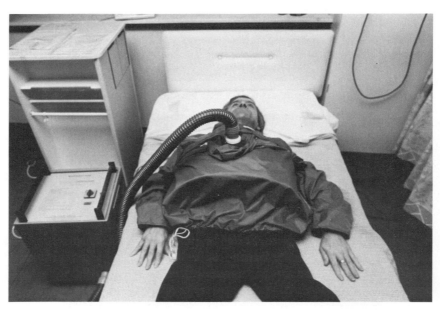

Fig. 4.14 Finished jacket respirator.

Table 4.5 Making a cuirass respirator

1	Make two simple Stockinette vests and put on the patient
2	Position the patient on a couch, lying as flat as they can
3	Using wide strips of plaster, make a cast of the patient's thorax, including as much as possible of the chest and abdomen from the sternal angle to the iliac crests and extending down on to the sides of the bed laterally (Fig. 4.10)
4	Remove the cast from the patient, cutting through the Stockinette vests
5	Build up the cuirass by 5–10 cm in the centre to allow the ribcage and abdomen to expand, but not at the edges which must ultimately form a good seal with the patient. (Compare Fig. 4.15 with Fig. 4.11.) Foam rubber or other material can be used to build up the cast, with a top layer of plaster.
6	Using a sheet of suitable thermolabile plastic (e.g. Fractomed), make a mould from the plaster cast. When hard, trim back the edges to conform to those of the cast
7	Cut strips of 2 inch thick PudgeeFoam, approximately 3 inches wide, and stick these to the inside rim of the top, bottom, and edges of the cuirass to provide an airtight seal
8	Make a strap from Neoprene. One edge of the strap can be glued to the cuirass shell, with Velcro for the fastening on the other edge. An alternative is to use Velcro on both edges of the strap, which the patient places on the bed before lying down, then positioning the cuirass shell and attaching the strap
9	Insert a connector (Fig. 4.13) to complete the cuirass (Fig. 4.16)

the frame and take the foam tubing off the edges. Use a hairdryer to heat the plastic and then mould it as necessary. Then reattach the foam.

Cuirass respirators

Cuirass respirators are also available commercially in a variety of different styles and sizes. While these may be adequate for patients without chest wall deformity, for other patients it is preferable to make an individual cuirass for them. As with jacket respirators, the materials and equipment are inexpensive and can be found in most occupational therapy or physiotherapy departments. A method for making a cuirass is described in Table 4.5 and illustrated in Figs 4.15–4.16. Higher pressures can be achieved than with jacket respirators, but normally a peak pressure of 15–30 cm H_2O below atmospheric pressure will be required. Small air leaks can be reduced by placing padding under the rim of the cuirass, while larger leaks may require adjustments of the foam seals, or

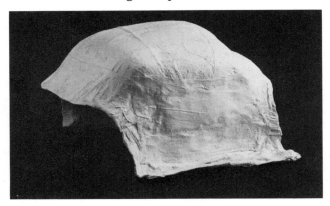

Fig. 4.15 Cast built up to act as a mould for a cuirass respirator. (Compare with Fig. 4.11.)

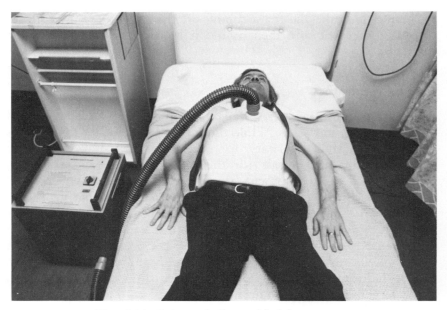

Fig. 4.16 Cuirass shell moulded from cast.

remoulding of the shell using a hairdryer to heat the edge of the plastic shell.

WEANING FROM t-IPPV

ENPV, like n-IPPV, can be used to wean patients from IPPV via an endotracheal tube. Sedation can be stopped and periods without assisted ventilation can be gradually increased. As with n-IPPV, certain criteria must be fulfilled (Table 4.6).

While removal of the endotracheal tube may be associated with a reduction in the volume of respiratory tract secretions, patients who are producing large volumes of sputum are probably better weaned by performing a tracheostomy. Since the airway will be unprotected, it is essential to ensure that the stomach is empty and will remain empty for the first few hours after the endotracheal tube is removed. Adequate function of the gastrointestinal tract must have been demonstrated by absorption of fluid administered via a nasogastric tube (in intubated patients, the absence of bowel sounds does not necessarily mean that the gut is non-functional). No fluid should be put down the nasogastric tube for four hours prior to extubation.

The patient should be placed in a cabinet respirator while still intubated (Table 4.7). Assisted ventilation should be transferred from IPPV to ENPV with the endotracheal tube still in place, and gas exchange monitored to ensure that adequate ventilation and oxygenation are achieved with ENPV. Immediately prior to extubation any secretions should be aspirated from the stomach, pharynx, and lower respiratory tract. ENPV should then be discontinued, the tube removed, and further pharyngeal aspiration performed before ENPV is recommenced. This routine will minimize aspiration.

Table 4.6 Prerequisites to using ENPV to wean patients from t-IPPV

Failed trial of conventional weaning techniques
Ability to sustain spontaneous ventilation for 15 minutes
Intact cough reflex
Minimal sputum production
Functional gastrointestinal tract
Empty stomach
Stable cardiovascular system
Low supplementary oxygen requirements ($FiO_2 < 0.4$)

Table 4.7 Using ENPV to wean patients from t-IPPV

1	Place in cabinet respirator with endotracheal tube in place
2	Transfer from t-IPPV to ENPV
3	Check gas exchange
4	Aspirate nasogastric tube
5	Aspirate pharynx
6	Pass suction catheter down endotracheal tube
7	Discontinue ENPV
8	Remove endotracheal tube
9	Aspirate pharynx
10	Recommence ENPV

COMPLICATIONS

The most common complication of ENPV is skin abrasion (Table 4.8). Soreness around the neck with a cabinet respirator can be overcome by the use of a soft foam collar. Skin problems over bony prominences with cuirass respirators are better treated by remoulding of the cuirass. Coldness due to air leakage can be overcome with additional padding or alterations to the seals to close the holes.

Table 4.8 Complications of ENPV

Skin abrasions (cabinet and cuirass respirators)
Upper airway obstruction
Pressure sores

UPPER AIRWAY OBSTRUCTION

If negative pressure is applied to the thorax at a time when upper airway muscle tone is reduced, partial or complete upper airway obstruction may occur. It is important to observe carefully for this phenomenon if external negative pressure ventilation is failing to maintain adequate gas exchange, particularly during sleep. Application of CPAP through a nasal mask has been used in combination with external negative pressure ventilation to overcome upper airway obstruction, but if a nasal mask is to be used it makes more sense to dispense with the external negative pressure device and change to n-IPPV.

CONTRAINDICATIONS

ENPV should not be used if the patient cannot protect their airway (Table 4.9). Making an effective cuirass can be difficult in patients with severe thoracic deformity. Unless gross, obesity is not a contraindication to ENPV, although it can create practical problems. The possibility of upper airway obstruction must be excluded in obese patients. Poor limb muscle strength or mobility may make it difficult for the patient to get into a cabinet respirator or to fit a jacket or cuirass, unless they are able to call upon assistance from family or carers.

Table 4.9 Contraindications to ENPV

Ventilator dependence for >15 hours a day (cuirass and jacket)
Gross obesity
Short neck (cabinet)
Tracheostomy (cabinet and jacket)
Severe thoracic deformity (cuirass)
Severe peripheral muscle weakness and/or poor mobility

MONITORING

As with n-IPPV, the intensity of monitoring during ENPV will obviously depend on the context in which it is used. Ideally, transcutaneous carbon dioxide tension should be monitored continuously to ensure that alveolar ventilation is adequate, with oximetry to assess oxygenation. Even then, arterial samples are valuable to monitor changes in acid—base balance, progressive resolution of base excess being a reassuring feature of effective ventilation. In practice, if the patient is well oxygenated during ENPV with minimal supplementary oxygen, alveolar ventilation is likely to be adequate and monitoring of carbon dioxide is not essential. (An important exception to this guideline is when supplementary oxygen is being administered, since a small increase in FiO_2 will abolish hypoventilation-induced hypoxia, so that SaO_2 may be normal despite significant alveolar hypoventilation.) Transcutaneous CO_2 electrodes are not as widely available as oximeters, but oximetry with periodic arterial blood gas sampling is a satisfactory alternative for most patients.

Cabinet respirators are particularly effective at providing assisted ventilation, and hyperventilation of the patient can occur. Arterial sampling is difficult in this situation, but a transcutaneous CO_2 electrode can be connected through one of the access ports in the cabinet.

The levels of oxygen and carbon dioxide during ENPV which are deemed satisfactory will depend upon the underlying respiratory condition. Irrespective of the initial level, a progressive fall in $PaCO_2$ indicates that ENPV is likely to be effective. Persisting hypoxia and hypercapnia may be improved by increasing the peak external negative pressure. Hypoxia without hypercapnia may necessitate addition of oxygen to the inspired air, but often improves spontaneously as secretions are cleared and atelectasis resolves.

After establishing that ENPV is producing adequate gas exchange, the frequency with which reassessments are made will depend on clinical progress. A daily arterial sample will often be sufficient. Non-invasive monitoring during sleep is essential to detect hypoxia or hypercapnia. Periods lasting 20–30 minutes when there is a slight deterioration in gas exchange may be seen, particularly during the latter half of the night. These are related to periods of rapid eye movement sleep, probably associated with a fall in lung volume and change in the pattern of spontaneous respiratory muscle activity. If the clinical condition of the patient is improving, no action is necessary, but a repeat sleep study should be performed a few weeks later.

TRIGGERING

Although technically possible, patient triggering of external negative pressure is seldom effective. For this, an additional sensor must be fitted to detect airway pressure, chest wall motion, or inspiratory muscle activity. The high flow rates during external negative pressure ventilation mean that there is a considerable time lapse after the trigger stimulus before sufficient negative pressure is generated to take over the work of breathing. Most patients settle comfortably after a short time on fixed rate ENPV, the distress caused by taking a breath out of synchronization with the pump being much less than with n-IPPV.

EXTERNAL POSITIVE PRESSURE

Application of external positive pressure can be used to assist expiration with cuirass, jacket, or cabinet respirators, but its value is very limited. The increase in tidal volume is generally small, and reducing the end-expiratory lung volume may actually decrease oxygenation as a result of shunting through atelectatic lung. With a cuirass, positive pressure tends to lift the shell away from the patient, making leaks worse. Positive pressure will distend a jacket and this can also be uncomfortable and make leaks worse.

CONTINUOUS NEGATIVE EXTERNAL PRESSURE

Continuous negative external pressure increases lung volume and improves oxygenation. If cyclical ENPV is superimposed on a background negative pressure, analagous to the use of PEEP in positive pressure ventilation, peak external pressures can be reduced. This technique can be beneficial in patients with neuromuscular disease and small lungs, probably by preventing the development of atelectasis.

5 Other assisted ventilation techniques

PHRENIC NERVE PACING

Implanted cardiac pacemakers are widely used to correct bradycardias, and a similar approach can be used to stimulate respiration in patients with inadequate ventilation. Electrodes can be implanted on the phrenic nerves and connected to a subcutaneous pacing box (Fig. 5.1), provided that the phrenic nerve distal to the electrodes is intact and the diaphragmatic muscle is normal. Phrenic nerve pacing may not produce adequate ventilation if the work of breathing is appreciably increased, for instance in chest wall deformity or airflow obstruction (Table 5.1).

The commoner situations in which ventilatory failure occurs with relatively normal respiratory mechanics, normal diaphragmatic muscle, and intact phrenic nerves, are central alveolar hypoventilation and high cervical cord transection. Before referring a patient for consideration for phrenic nerve pacing, normal phrenic nerve conduction times should be demonstrated by transcutaneous stimulation in the neck. Figure 5.2 shows the use of electrodes to stimulate the phrenic nerve, the resultant muscle action potential being recorded with electrodes in the seventh and eighth intercostal spaces. The stimulating electrodes are positioned anterior to the sternomastoid muscle, but should be moved inferiorly and anteriorly in the anterior triangle if the phrenic nerve cannot be found. Typical stimulation parameters would be a pulse width of 100 microseconds at 100 volts. When the nerve is stimulated, a diaphragmatic twitch will be visible on the abdomen and the recording electrodes will register an action potential. The latency from the stimulus to the onset of the action potential should be less than 10 milliseconds (Fig. 5.3).

Table 5.1 Prerequisites to referring a patient for phrenic nerve pacing

---
No clinical suggestion of neuropathic or myopathic process
No severe chest wall deformity
No airflow obstruction
No obesity
Phrenic nerve conduction time normal

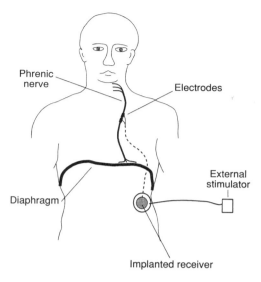

Fig. 5.1 Phrenic nerve pacing.

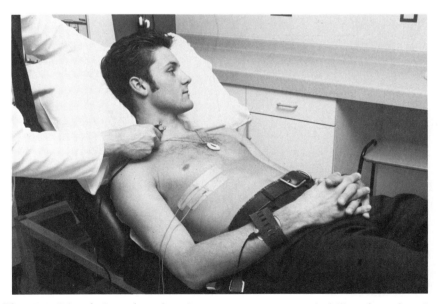

Fig. 5.2 Stimulating the phrenic nerve to assess suitability for phrenic nerve pacing.

Continuous stimulation of one phrenic nerve for long periods can damage the nerve and the diaphragm. If adequate ventilation can be produced by stimulation of one phrenic nerve, implantation of bilateral systems will allow continuous ventilatory support by stimulation of each nerve for alternating periods. Alternatively, phrenic nerve pacing can be alternated with another method of assisted ventilation, usually t-IPPV. This strategy will also be necessary in patients who require simultaneous bilateral phrenic nerve stimulation for adequate ventilation.

Since upper airway muscle contraction is not coordinated during phrenic nerve stimulation, upper airway obstruction may occur, particularly during sleep. CPAP or a tracheostomy may then be necessary. The external stimulator equipment for phrenic nerve pacing is compact, and facilitates day-time mobility in patients who are completely dependent on assisted ventilation. Nevertheless, if a tracheostomy is necessary because of upper airway obstruction or for any other reason, the cost of a phrenic nerve pacing system may not be justified and t-IPPV may be a more satisfactory alternative.

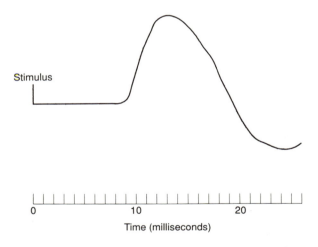

Fig. 5.3 Action potential recorded from the diaphragm during stimulation in Fig. 5.2.

ROCKING BED

The weight of the contents of the abdominal cavity leads to postural changes in respiratory mechanics. On changing from the upright

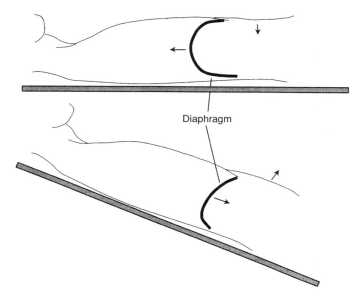

Diaphragm

Fig. 5.4 Principle of action of the rocking bed.

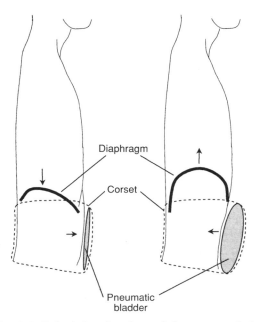

Diaphragm

Corset

Pneumatic
bladder

Fig. 5.5 Principle of action of the pneumobelt.

to the supine posture, the tone in the diaphragm must increase to prevent the abdominal contents from rising into the chest. In a rocking bed, the effect of gravity on the abdomen is used to provide a piston-like action which inflates and deflates the lungs. The bed rocks between horizontal and head-up positions. Changing from horizontal to head-up causes the abdomen to descend, drawing the diaphragm down with it and thus inflating the lungs. Conversely, returning to the horizontal results in expiration (Fig. 5.4).

Obviously the airway is not protected in a rocking bed, so it is not suitable for patients who are in danger of aspiration. It does not produce effective ventilation in patients with a stiff chest wall, airflow obstruction, or if the work of breathing is increased for any other reason. In patients with normal respiratory mechanics, for example with an isolated diaphragm paralysis, it can be a simple and effective treatment for nocturnal hypoventilation.

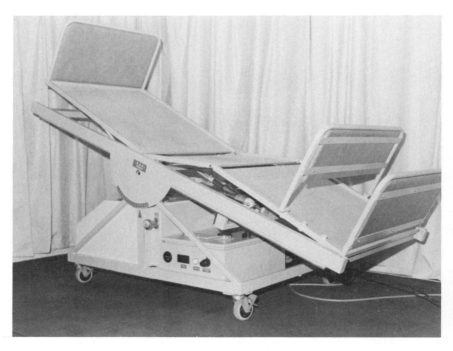

Fig. 5.6 Rocking bed.

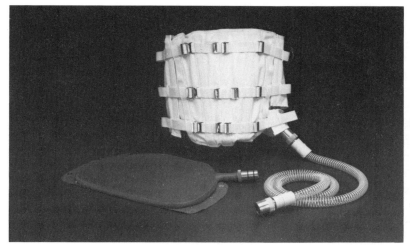

Fig. 5.7 Pneumobelt.

PNEUMOBELT

The pneumobelt also utilizes the effect of gravity on the weight of the abdominal contents (see previous section). With the patient sitting upright, expiration is achieved by inflating a pneumatic belt, around the abdomen, forcing its contents up into the chest. On releasing the pressure in the belt, the abdomen descends under the influence of gravity, and air flows into the lungs (Fig. 5.5).

Pneumobelts, like rocking beds, are not a particularly effective means of assisting ventilation. In patients with relatively normal respiratory mechanics who are almost totally dependent on some form of assisted ventilation, they can be used during the day-time in a chair or wheelchair, reverting to more effective ventilation such as t-IPPV at night.

HIGH FREQUENCY CHEST WALL OSCILLATION

Ventilation at frequencies above the normal breathing rate leads to oscillation of the air within the lungs, improving gas exchange. High frequency ventilation or oscillation has been extensively investigated in the intensive care setting using an endotracheal tube, but oscillating the air around the chest wall using a rigid cuirass shell produces similar effects. The place of this technique in providing assisted ventilation at home has yet to be established.

6 Assisted ventilation of children at home

BACKGROUND

Many children were among those affected by the poliomyelitis epidemics earlier this century. Only a small proportion of these developed respiratory failure, and many of those that did needed ventilatory support for only a limited period. Nevertheless, some required long term nocturnal or continuous assisted ventilation. Most of these were ultimately transferred to their own homes, where they used a variety of positive and negative pressure devices. Poliomyelitis is now rare, but the experience gained with these patients has led to the concept of assisted ventilation at home being extended to children with ventilatory failure from other causes.

SELECTION OF CHILDREN

Selection of children for assisted ventilation depends upon the underlying disease process, the clinical condition of the child, the family, and the home environment.

The underlying disease

Children who may be considered for assisted ventilation at home fall into several distinct groups. Table 6.1 lists the conditions affecting children in which assisted ventilation at home may be considered. Idiopathic central alveolar hypoventilation usually presents soon

Table 6.1 Conditions in which assisted ventilation at home in children may be considered

---
Central alveolar hypoventilation syndromes
Cervical cord or phrenic nerve injury
Neuromuscular diseases with disproportionately severe respiratory muscle involvement
Infants with congenital abnormalities or immature lungs who cannot be weaned from assisted ventilation

---

after birth, although occasionally diagnosis may be delayed for several years. Apnoeic episodes or cyanosis are apparent during sleep, but if these pass unnoticed the first presentation may be with right heart failure secondary to nocturnal hypoxia. Congenital or acquired brainstem abnormalities may be associated with a similar clinical picture, which can also develop after encephalitis. In the absence of other problems the prognosis with assisted ventilation is good. Since assisted ventilation is only needed during sleep, the practicalities of assisted ventilation become easier as the child grows older and the amount of time spent asleep declines.

Children who become quadriplegic as a result of trauma to the cervical cord will develop respiratory failure if the level of damage is above the mid-cervical region where the phrenic motor neurones originate. Diaphragmatic paralysis can also develop after thoracic surgery with inadvertent damage to the phrenic nerves.

Progressive neuromuscular diseases may present in infancy, but more usually respiratory problems are first encountered when the child is several years old. Assisted ventilation is undoubtedly of value for children who have relatively mild generalized muscular weakness, with disproportionately severe respiratory muscle weakness. Nocturnal assisted ventilation will abolish day-time symptoms caused by hypoventilation during sleep, and will also reduce the incidence of life-threatening episodes of acute respiratory failure. These children can gain many extra years of good quality life.

The role of assisted ventilation with more severe generalized neuromuscular problems is more controversial, and the pattern of use differs both between and within different countries. This is discussed in more detail in Chapter 1. Life can certainly be prolonged, but the question arises as to whether this is of sufficient quality. Assisted ventilation can be an unwelcome additional imposition on the child and family. Nevertheless, a good quality of life is undoubtedly possible despite being wheelchair bound and ventilator dependent. Anticipation of the need to make a decision on assisted ventilation is important for these children and their families (see section on the family below).

The advances in neonatal medicine in recent years have meant that an increasing number of infants with congenital abnormalities or immature lungs survive. A proportion of these cannot subsequently be weaned from assisted ventilation. Although the prognosis for these children is less good than for those with central alveolar hypoventilation, and the demands of home assisted ventilation considerably greater, some families prefer to look after the child at home.

The child's clinical condition

Criteria which must be fulfilled before a child is discharged from hospital with assisted ventilation are given in Table 6.2. It is essential that the child is medically stable before discharge from hospital, with low supplementary oxygen requirements. Children who have right heart failure will not do well on assisted ventilation at home, and every effort should be made to control heart failure prior to discharge. Adequate means of nutrition should be established, with a clearly documented positive growth trend.

The family

A supportive family keen to look after their child at home is central to the success of assisted ventilation of children at home. The decision to take on this responsibility can be difficult for the family, with implications for their relationships with each other, mobility, social life, and financial status. It is important that they are given as much warning as possible of the need to make this decision. Serial measurements of vital capacity and $PaCO_2$ will often give several months' warning of the need for assisted ventilation, and this period should be used to discuss the question with the child and family. It must be made clear to them that while assisted ventilation may prolong life, the underlying disease will continue to progress. In a child with progressive neuromuscular disease, starting assisted ventilation may cause the family to retain hopes of a cure and deny the inevitability of later decline.

Table 6.2 Criteria for assisted ventilation of children at home

---

Stable medical condition
Infrequent lower respiratory tract infections
No right or left heart failure
Low supplementary oxygen requirements
Safe and effective assisted ventilation established
Adequate nutrition
Positive growth trend
Safe home environment
Family keen to have the child at home
Family or carers competent to care for child
Satisfactory hospital backup services

Families are often aware, through support groups and other channels, that assisted ventilation is an option for their child. Even when the hospital team feel that assisted ventilation at home would clearly be inappropriate, it is often prudent to broach the subject with the family to ensure that an informed decision is made which is satisfactory to all concerned. Withdrawing assisted ventilation is difficult for all concerned once it has been started, and the temptation to put off this difficult decision must be avoided. It may help them in their decision to talk to other families with a child using assisted ventilation at home. During discussion it may become apparent that the family do not feel that assisted ventilation is appropriate for their child, or an older child may make their own decision that they do not want to take this step. If this is clearly the conclusion at which the family is going to arrive, then the responsibility for taking it should be assumed by the physicians caring for the child, to alleviate any guilt on the family's part.

The home environment

In addition to the proficiency of the family, a safe home environment is also essential. The home should be sufficiently spacious to accommodate all the equipment needed by the child. Some alterations may be necessary, for instance to create wheelchair access, a storage area for backup supplies, or a clean area for tracheostomy care. Nevertheless, the atmosphere of a home should be preserved, without changing it into an outpost of the intensive care unit. The aim should be to integrate the child not just into their family and home, but also into the community and friends. This aspect will need encouragement and support from the hospital and community carers, to give the family confidence to bring friends and neighbours into the home and to take their child out into the community. If the child can attend a local school, staff and pupils will need to be educated about assisted ventilation.

ASSISTED VENTILATION TECHNIQUES IN CHILDREN

n-IPPV

The place of n-IPPV for ventilating children at home is evolving, but it is likely to become the most commonly used technique. Small

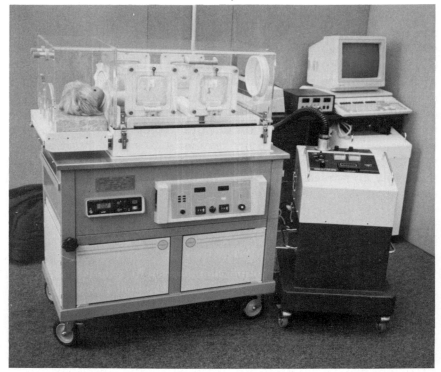

Fig. 6.1 Paediatric negative pressure chamber and ventilator.

masks for paediatric use are now available from several manufacturers. Children under the age of five years do not tolerate nasal masks well, and other techniques are better suited to children below this age. The same ventilators are used as in adults, and commencing a child on n-IPPV follows the same principles as outlined in Chapter 2.

t-IPPV

Children with severe neuromuscular disease often need assisted ventilation for most of the day and night. At this stage in their disease they have usually lost the ability to cough effectively and may be unable to protect their airway. Tracheostomy and IPPV remains the optimal mode of ventilation for these children. As with n-IPPV, the equipment and ventilators are the same as those used in adults (see Chapter 3).

ENPV

Paediatric sizes of cuirass, jacket, and cabinet respirators have been available for many years. They were designed initially for poliomyelitis victims, but more recently have been used for children with neuromuscular diseases. Smaller versions of conventional cabinet respirators can be used in older children, but in infants the high air flow rates can lead to problems with temperature control. Figure 6.1 shows a negative pressure ventilator developed for younger children. This differs from the conventional design by having a heater unit situated underneath the chamber, and by using a rubber neck seal which is placed on the child before they are put in the cabinet. This seal can quickly be detached from the cabinet to remove the child.

In addition to avoiding a mask over the face, the dexterity required for a child to fit their own cuirass is less than for a nasal mask, and children often prefer ENPV to n-IPPV for this reason. The principles of making a personalized cuirass or jacket are identical to those outlined for adults.

Phrenic nerve pacing

Phrenic nerve pacing has been used successfully in children, although simultaneous bilateral pacing is usually necessary to obtain adequate ventilation. This is probably a result of the relative ease, compared to adults, with which the ribcage distorts during diaphragmatic contraction.

7 From hospital to home

INTRODUCTION

The success of any home ventilation programme depends just as much on the way patients are transferred back to the home environment as it does on the technical aspects of establishing assisted ventilation. This chapter outlines the general principles involved, which can be interpreted for each individual patient and their own particular abilities and home.

DISCHARGE PLANNING

As soon as it is apparent that a patient is likely to be established successfully on assisted ventilation, a discharge plan should be formed. Whenever possible, this should be the responsibility of one member of the health care team, with whom the patient can discuss any particular problems or anxieties. This person should also accompany the patient when they first leave hospital. If the patient needs care from family or community services, these individuals must be taught the necessary skills and be involved in setting a discharge date.

The first step in getting a patient accustomed to being out of a hospital environment is to take increasingly long trips around the hospital and its grounds. This is particularly important for ventilator-dependent patients (see below). A checklist of all equipment that will be needed should be made, together with written action plans to cover emergencies, ventilator failure, and power cuts (Table 7.1). The patient and their carers should be told how to identify symptoms of inadequate gas exchange or chest infections, again with a written plan of action. If appropriate, the carers should be instructed in cardiopulmonary resuscitation.

THE HOME ENVIRONMENT

For some patients, for example those who are only going to use n-IPPV or a cuirass at night, discussion with the patient and their family may suggest that a home visit before discharge will not be necessary. The home can be assessed by the member of the team who

Table 7.1 Example of action plan for patient using n-IPPV at home

Name: D. A.
Hospital record number: N123456
Ventilator make: Nippy
Serial number:
UHN medical equipment service unit number:

Contact telephone number: Nottingham 421421, extension 44470
Emergency number: Nottingham 421400, extension 43712
General practitioner's number: Nottingham 123456

If your ventilator alarm sounds:
If the ventilator is not working at all, check that it is plugged in and that the mains power is switched on. If the alarm still sounds, contact the ward. If the alarm sounds while the ventilator is working, look and see whether the red lights on the top of the ventilator control panel are lit up. It is unlikely that the 'High' light will alarm, but 'Low' will light if the tubing is disconnected or if your mask is leaking. Check these, and if the alarm still sounds contact the ward.

If your ventilator stops working altogether:
Check that the mains lead is connected and plugged in, and that the plug is turned on. Check that the ventilator is turned on. If it still does not work, contact the ward and we will ask you to bring the ventilator in or send a replacement out to you.

If there is a power cut:
The local emergency services are aware that you have a ventilator which works on electricity, and will reconnect you as a matter of high priority. If possible, telephone them on Nottingham 654321 to check that they know you have been cut off. Your ventilator will not work from batteries. If after 2 hours the electricity supply has not been reconnected, telephone the ward (number given above) and come in to hospital. The ward will arrange transport for you if necessary.

If you develop a chest infection:
If your phlegm changes colour or if you develop a temperature, take your reserve antibiotics, one tablet four times per day. If after 2 days you are no better, call your general practitioner for some stronger antibiotics. We have let your general practitioner know what to do if you are not getting better. If you cannot get hold of your general practitioner, or if you feel at any stage you are getting worse, call the ward. Remember to get another reserve supply of antibiotics when you get over the infection.

Ventilator settings (see photocopy of controls):

Pressure—about 11 o'clock on the dial	Trigger—1
Low pressure alarm—10	High pressure alarm—25
Insp—1.5	Max Exp—2.5

accompanies the patient when they leave hospital. For the majority of patients, a trial visit home for a few hours will be useful to identify any practical problems which require attention. There must be an electric socket close to the bed where the ventilator will be used, and for ventilator-dependent patients additional electricity sockets should be fitted so that the ventilator, suction apparatus, wheelchair battery recharger, and so on can all be plugged simultaneously directly into wall sockets in the areas where the patient will sleep and spend the day-time. If the patient is confined to a wheelchair, doors may need to be widened and ramps fitted. Toilet facilites may require alterations and a stair lift may also be needed.

THE FIRST TRIP HOME

Discharge should take place in the morning, so that any problems that arise can be dealt with later that day. The first night at home is often extremely stressful for the patient and their carers, and the ideal solution is to have a member of the hospital team sleep in the same house for the first night. If this is not possible, a competent member of the family or other carer should be resident that night, and the home should be visited by one of the team the next day, to sort out any problems, discuss anxieties, and arrange readmission to hospital that day if necessary.

Community services and the patient's own doctor should be informed of the discharge. Local services such as the fire brigade should also be informed, so that emergency electricity supplies can be established by them in the event of a power cut.

READMISSION ARRANGEMENTS

At the time of initial discharge, a clear contact route must be identified which guarantees automatic readmission at the patient's request. This should only require one telephone call by the patient to a ward that is staffed for 24 hours a day. The staff on the ward should have details of all local patients on home ventilation, and have the authority to organize emergency transport.

The timing of routine follow-up visits will vary for each patient. Normally they should be seen within one month of discharge and every three months thereafter. Depending upon the distance the patient lives from the hospital base, these reassessments may be short readmissions, out-patient clinic visits, or visits by members of the team to the patient's home.

Planned respite admissions should be arranged for the more dependent patients and children, in order to give the carers some relief. The timing and duration of these will vary, but an admission to hospital for one week every three months or so may greatly enhance the ability of family to cope with assisted ventilation at home.

MONITORING

When the patient attends hospital, arterial blood gases should be checked. In patients with nocturnal hypoventilation, day-time respiratory failure usually continues to improve for several months after commencing assisted ventilation (Fig. 7.1). A full blood count is also useful to detect polycythaemia, indicating that oxygenation may be inadequate. Serum electrolytes should be measured regularly, particularly in patients who require diuretic therapy for right heart failure. Routine sleep studies are not necessary if day-time blood gases are improving or stable. Inadequate ventilation at night will usually lead to recurrence of day-time symptoms, deterioration in day-time arterial blood gases, and right heart failure. A sleep study should then be performed. Routine reassessments may also include measurement

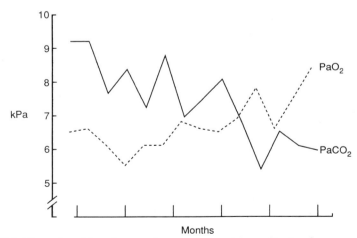

Fig. 7.1 Day-time blood gases in a patient with scoliosis after commencing nocturnal assisted ventilation.

of vital capacity, maximum mouth pressures, and so on, but these seldom lead to changes in management. Table 7.2 shows an example of a monitoring chart.

MAINTENANCE

Prior to discharge, a maintenance programme must be drawn up (Table 7.3). Arrangements must be made for regular supplies of disposable items. A patient using n-IPPV may only need a small filter every few months and a new mask each year, whereas a patient with a tracheostomy will require regular supplies of dressings, suction catheters, and so on. Each ventilator has its own maintenance programme, and it is essential that this is adhered to, with a replacement unit being provided if necessary while servicing is carried out. If the ventilator company has a service contract then the work may be done at home; otherwise, hospital personnel can service the unit coinciding with a planned readmission or clinic visit (Table 7.2).

Table 7.2 Example of monitoring chart

Name:
Address:
Telephone number:
Hospital number:

Date commenced home ventilation: 5/12/91
Ventilator make: Nippy
Serial number:
UHN medical equipment service unit number:

Date	Ventilator service Y/N	Sleep study Y/N	$PaCO_2$	PaO_2	Comments
6/1/92	N	Y	7.7	8.3	Coping well
12/3/92	N	N	6.8	8.0	No problems
2/5/92	N	Y	7.2	7.3	More breathless; for sleep study
10/5/92	N	Y	6.6	7.7	Mouth opening; SaO_2 now fine with chin strap
16/8/92	N	N	6.1	8.8	Well
5/10/92	N	N	6.3	9.1	Well; reduce diuretics
15/12/92	Y	N	5.9	9.3	Well; FBC, U&E

Table 7.3 Example of maintenance schedule for patient using a Nippy for n-IPPV

Each day
Check ventilator connections
Check that ports on mask closed

Each week
Check ventilator settings
Wash, dry, and replace ventilator filter
Wash and dry mask

Each month
Wash and dry ventilator tubing
Wipe outside of ventilator with dry cloth

Every year
Replace mask
Replace headgear
Check tubing
Electrical safety check on ventilator
Read ventilator clock

Ventilator tubing and masks need to be cleaned weekly, and care taken that they are allowed to dry completely. Filters should also be checked weekly. The controls on some ventilators can accidently be changed quite easily, and their settings should also be checked each week.

THE VENTILATOR-DEPENDENT PATIENT

A patient who cannot breathe spontaneously, for instance following high cervical cord transection, is a considerable challenge to care for at home. In many cases, particularly those patients with cervical cord transection, initial care is best provided by a specialist unit such as a spinal injuries unit, whose staff are skilled in helping with the multiple problems which the patient has to face.

Table 7.4 lists the criteria which must be satisfied for a ventilator-dependent patient to be considered for transfer to their home. Most ventilator-dependent patients will use t-IPPV and the technical requirements for this mode of ventilation are described in Chapter 3.

Table 7.4 Home discharge criteria for ventilator-dependent patient

---

Stable medical condition
Stable gas exchange on ventilator
Inspired oxygen less than 40%
Adequate nutrition established
Manageable secretions
Stable cardiac rhythm
No other major change in management anticipated
Motivated patient
Motivated and capable carers

Caring for a ventilator-dependent patient at home is a considerable responsibility. Those who are going to care for the patient at home must understand the full implications for them, and be highly motivated to undertake this challenge. Their learning abilities must be assessed carefully, and then a training programme can be planned for them. During this training period it may become apparent that, despite encouragement, their motivation is not as high as was initially thought and alternative plans must be made. Strong vocal support does not necessarily mean commitment, and a prospective carer's worries about the responsibility for home ventilation may be manifest as inability to learn tasks in the build up to discharge. The need of family carers for professional support will vary depending on the capabilities of the family and the needs of the patient, and it is essential to ensure that community services are willing and able to contribute to the care of the patient. Obviously, professional carers who will be involved in home care must be involved in the training programme along with the family. The training programme must cover all aspects of the patient's care, progressing over a period of several weeks from simpler tasks such as connecting up the ventilator to more complex procedures such as changing the tracheostomy tube and checking the cuff volume (Table 7.5). Table 7.6 outlines the equipment which will be needed.

Table 7.5 Training programme for carer of ventilator-dependent patient

1	The underlying disease process
2	The basic principles of assisted ventilation
3	Administration of routine medication
4	Monitoring fluid balance
5	Maintenance of adequate nutrition
6	Moving patient around the home
7	Chest physiotherapy
8	Setting each ventilator control
9	Ventilator alarms
10	Connecting external battery to ventilator
11	Emptying water from ventilator tubing
12	Tracheostomy care:
	Aspiration through the tracheostomy
	Instillation of saline
	Dressing the stoma
	Sterilization of spare tube
	Inserting an inner tube
	Changing the tracheostomy tube
	Checking cuff volume
	Use of tracheostomy dilators
	Emergency insertion of a smaller tube
13	Recognition of deterioration in clinical condition
14	Cardiopulmonary resuscitation
15	Emergency procedures:
	Ventilator/power failure
	Blockage of tracheostomy
	Displacement of tracheostomy tube
	Leaking tracheostomy tube cuff
	Cardiac arrest
16	Maintenance
	Cleaning and replacing filters
	Disconnecting, cleaning, filling, and replacing humidifiers
	Changing and cleaning ventilator tubing
	Ventilator internal battery check
	External battery check

Table 7.6 Equipment for ventilator-dependent patient at home

Apparatus for t-IPPV at home
Main ventilator
  Spare ventilator tubing
Tracheostomy care disposables
  Tracheostomy tape
  Tracheostomy stoma dressings
  Bandage scissors
  Wound cleansing fluid
  Spare tracheostomy cannula
  Sterilizing fluid
  Small tracheostomy cannula for emergency use
  Tracheostomy dilators
Humidifier
  Sterile water
Suction apparatus
  Mains-powered suction unit
  Battery-powered back-up suction unit
  Suction catheters
  Syringes
  Saline
  Gloves

Mobile t-IPPV
External battery for ventilator (separate from that operating wheelchair)
Battery-operated suction
Regeneration humidifier

Monitoring
Airway pressure monitor
Remote alarm

Backup ventilation
Self-inflating bag
Additional external battery or generator
Battery charger
(Spare ventilator)

8 A patient's view of living with a ventilator—Harry Smith

INTRODUCTION

This chapter is bound to be somewhat autobiographical, not because I am famous or particularly special in any way, but because if I am to add anything at all to this book, I need to relate my personal experiences and emotions. Nearly forty years ago I had poliomyelitis, in the year that preceded the introduction of the Salk vaccine into Britain, and in 1988 the illness tried to catch me again through the surprisingly common condition of apnoea. As is detailed elsewhere in this book, apnoea can be caused by a range of underlying problems; in my case it was the respiratory weakness beqeathed to me by polio that brought the sleep problems in later life. The insidious way in which the problems developed over many years, and how my second rebirth has been made possible, is the crux of my message to people with similar sleep difficulties, and to those charged with diagnosing and treating them. I hope to be able to encourage those who find they have to live with a ventilator by telling how it has given back my life in full measure, how I have travelled to many parts of the world with it—I am writing this in St Lucia in the Caribbean—and how best to get used to the ways in which the machine forces changes in one's activities.

THE FIRST REBIRTH

In 1956, I was 20 years old, a fitness fanatic, full of adventure, a sporting cyclist, a hockey player, a rock climber, and a trainee pilot. I was a student at the end of my second year at the University of Manchester reading for a BSc in botany (the study of plants). I will not dwell on polio, as this chapter is about its long term effects. It is a strange disease, though. One Friday that year, I began to feel sick and queasy. By Saturday, having taken my girlfriend to the local hop, I could hardly drive my father's car back home, and on Sunday my mother called the doctor. My father had also been sick that week and the doctor reckoned it was summer flu. By Monday morning I felt a great deal better, but on waking I was initially amused to find that I could not move my left arm—I thought I must have slept on it awkwardly. Within a few minutes, however, I began to realize what was happening, and a couple of hours later was in an isolation hospital. In hospital, for a while I followed the progress of my paralysis in a strange, almost clinical detachment. During the Monday night, a nurse whose name has gone from my memory, but whose care and devotion will always be with me, held my hand for hours while my breathing slowly subsided and eventually stopped completely. It was about three weeks later that I regained a conscious appreciation of myself and my condition, and by that time the nurse had left the hospital.

I awoke from delirium to find myself locked into a box, like a coffin, but with my head sticking out from one end. Officially known to medics as a tank respirator, the rest of us call it an 'iron lung'. For a while I fought this monstrous, wheezing machine, but I was still very ill, now exacerbated by pneumonia. So, eventually, I learnt that this awful, fearsome machine, about which I still have nightmares, had saved my life and I grew to love it. Indeed, I grew to become totally dependent upon it. That I was able to break free at all was entirely due to the fortunate chance of the spread of the virus; some of those infected at that time were never able to resume a life without respiratory assistance, and I count myself as incredibly fortunate. After the infection burnt itself out and the medics started to assess the lasting damage, it was discovered that I could breathe, very faintly and with considerable difficulty, using only my diaphragm. I had also regained the use of my legs, which had temporarily given up the fight, and after about three months I was taken out of the respirator to try walking. I had my 21st birthday in the lung, surrounded by hundreds of cards and with visits from local dignitaries—I had become a nine-day wonder. It was in preparing for the party that it was discovered I could breathe with my diaphragm, and I was even allowed a glass of beer to celebrate. An event etched in my mind is the doctor's expression of sheer delight when he saw that I was breathing without the machine for the first time.

It took me a couple of years to recover from polio, as far as I ever would, and in that time my life was changed beyond recall. I could no longer cycle, play hockey, swim, fly airplanes, or climb rock faces, and at first I was mortified. But I could drive a car, and I could walk and even run at times. As a direct result of my illness I met again a girl I had known previously, and without whose devotion and support ever since I would have been useless. I went back to university, got my degree, then a PhD, and eventually I became a professor. In between, my new girlfriend agreed to marry me and we raised three splendid children. This is what I meant by rebirth. Everyone has watersheds in their lives; mine was particularly obvious, and it changed me from an uncommitted drifter into someone with drive and ambition. Whether this is an improvement or not I leave others to decide.

THE DECLINE

Thirty-one years later, I found myself in the intensive therapy unit of the City Hospital in Nottingham being ventilated once again, but this time via a tube stuck down my throat. This respiratory failure was the culmination of at least five years, and perhaps as much as fifteen years, of physical deterioration, the causes of which I could not fathom. The only way I can put any timings to what happened is by trying to identify symptoms that have disappeared since the condition has been controlled. Control is the correct word, since as I understand it once you have my sort of apnoea you do not get rid of it; all you can do is control it. My apnoea is controlled by the use of a ventilator at night, and for the five years in which I have had night-time ventilation, every one of the symptoms I shall now describe has vanished. So, what I shall now do is try to take the reader through 15 years of gradual, inexplicable deterioration.

Indigestion

On the basis that problems I had have now gone, I suspect the apnoea began about fifteen years before it was diagnosed, and the first symptom was, of all things, indigestion! I say this because I used to have terrible trouble after meals, and now I do not. The indigestion began slowly, but developed to a point at which I sought medical advice. I even had barium meals and X-rays to look for a suspected ulcer, but nothing was seen. This was the first of many occasions on which I was told it was all in my mind.

Irritability

Coupled with indigestion was increasing irritability, manifesting itself both at home and at work in short-temperedness and impatience with others. Since I have always had a short fuse, nobody thought this was anything other than an aggravation of an innate characteristic. It did not make anyone love me any the stronger, of course! Both the indigestion and the irritability became gradually worse over the 15 years or so, until I guess I became just about impossible to live or work with. That family and colleagues put up with me for so long continues to amaze me. Later, one of my closest friends and colleagues told me that, towards the end, I was in danger of having a complete mutiny on my hands; he had supported me by insisting that there must be something wrong with me, that it could not all be down to stress and overwork, as most thought. He seems to be the only person close enough, yet sufficiently detached, to have recognized that the problems were not just psychological. Since being ventilated both my wife and my colleagues say that I am considerably mellowed compared to pre-1988!

High blood pressure

Four or five years before my apnoea was diagnosed, I had to have a full medical examination for a life insurance policy. It turned out that my blood pressure was raised significantly, but not yet to the level that treatment was required. Only regular monitoring was called for, and every three months thereafter the doctors watched it slowly climbing. Accompanying the raised blood pressure was a banging in the ears, particularly at night-time. Within a couple of days of being ventilated my blood pressure fell to normal levels, and the ear thumping has completely disappeared.

Sleeplessness

I cannot now remember when I first began to worry about my sleeplessness. I had had for many years major difficulties in sleeping in strange beds, and if I was away from home for anything less than about three nights I might as well not have gone to bed at all. I had always put this down to a relatively common difficulty in accommodating to fresh surroundings, but even this stopped being a big issue after ventilation. Somewhere between 10 and 15 years before diagnosis, I began to be aware that I was waking up more often than I should, each time in order to urinate. The 'tinkle trot', as it became known within the family, became a regular and increasingly frequent feature of my life. In the last year before I started ventilation I would have to go to the loo at least five times, and often as many as ten times every night. At this stage, I became considerably disturbed, and insisted on tests for diabetes and an examination of the prostate. Both turned out to be negative and my doctors came to the conclusion that stress was the root of all my problems. Like all the other symptoms, since ventilation the 'tinkle trots' have gone, and I can sleep all night through, even in strange beds.

Day-time somnolence

Although the sleeplessness was worrying, what became even more frightening was what I now know as daytime somnolence—in other words, going to sleep during the day. The frightening part was not the drowsiness, which was almost always there, but the incredible suddenness with which my head would drop and I would fall asleep. It would happen at work, or at home. On one occasion my head crashed to the table in a meeting of a Royal Society committee of which I was (temporarily) a member; the embarrassment was acute, I can tell you. Although it is regarded as quite normal for members of University Senate to go to sleep at meetings, when I did it, it was with a crash! Coming home for dinner, I once dropped my head straight into my soup. If you have not experienced the abrupt, complete loss of control, you cannot appreciate the fright it gives you. This particular aspect of the condition reached major significance when, just before Christmas in 1987, I fell asleep at the wheel on the A46; it was fortunate that the A46 at that point was dual carriageway, or this story might not be being written.

Early morning headache

The last symptom of apnoea that I was to experience was early morning headache. This was not a dull pain, but a searing bullet-through-the-temple agony that greeted me every morning in the last few weeks before my respiratory breakdown. The nature of this experience convinced me yet again that I had a serious neurological problem.

CRISIS

The Christmas of 1987 was particularly grim for me and, I suspect, my family. I felt dreadful, having just missed killing myself by falling asleep at the wheel, and having virtually no sleep from one night to the next. I felt so bad that I went again to my doctors, and one of them prescribed sleeping pills. These did no good at all, and by January 3rd I was feeling like death. At about 10 o'clock that evening I took two sleeping pills and began to get ready for bed, but was interrupted by the arrival of our second daughter and her boyfriend with the announcement of their engagement. Naturally we were delighted and proceeded to celebrate. I had a large whiskey and we talked for a while and then went to bed. My wife woke at 7.30 to find me very, very cold and very, very blue in the bed by her side. She called the doctor and I was transported in an ambulance to the City Hospital in Nottingham.

THE INTENSIVE CARE UNIT

Intensive care is a very personal experience, and although I feel it is necessary to deal with my experience in this chapter, I have to draw some veils. During the first one or two days, I must have been totally uncommunicative to those who were caring for me, but I lived a life of my own. Even now, I cannot find the courage to describe in detail the hallucinations that filled my brain. After about two days I regained some contact with my surroundings, largely through the searing pain in my mouth and throat. I found that I had tubes in virtually every natural orifice, and one or two besides. The one in my mouth seemed enormous, blocking my throat and forcing my tongue down into the floor of my mouth. It was attached to a large gurgling and wheezing machine by what passed for a bed. In such a situation one is understandably frightened, and the worst part of it was not being able to communicate to anyone. My wife was there, looking confident and cheerful, so I guessed I must not be in too bad shape. I suspect anyone who has experienced intensive care has his or her own very individual memories. Mine included the constant aggravation of the radio. I also puzzled for several hours trying to decipher what was written on a vivid yellow garbage bag situated just beyond the end of my bed: clinical waste for incineration. In my state of heightened sensitivity I felt that the comment referred to me, not just to the waste from the ward, particularly in view of the regular clattering of the 'jolly trolley' as yet another departed was removed from the unit. I was intubated for 8 days, but thereafter I still needed 'sucking out' so a mini-tracheostomy was inserted. I knew the design of this because at the end of my bed there was a poster giving instructions on how to fit and replace one; apart from the clinical waste bags, this poster was the only thing I had to read for a week. My abiding memory, however, is of the most dedicated team of professionals I have ever had the privilege to witness. I never want to experience it again, but I regard my time in intensive care as one of the most humbling, yet rewarding experiences of my life.

THE SECOND REBIRTH

By chance, the first object that my eye focused on when I entered the assisted ventilation unit at Newmarket was an iron lung identical to one in which I had spent several months in 1956. My dismay can be imagined when I jumped to the conclusion that I was going to have to go in it once again. My agitation must have been apparent to Dr John Shneerson, who calmly explained to me about apnoea. Instantly, it was revealed to me that everything I had been suffering from for countless years could be traced to my strange habit of stopping breathing during the night. In one sense the relief was colossal, but it was truly terrifying to be told that I would never be able to sleep without breathing assistance again. A cuirass would not work for me, and I use a nasal mask ventilator. Today it is second nature to me to strap on the mask, connect up, and switch on, but at first it took three of us—me, my wife, and my son—to get me going at night.

LIVING WITH A VENTILATOR

Ventilators are machines, and machines have two principal characteristics—they make a noise and they can go wrong. These two characteristics are probably the most worrying aspects of living with a ventilator. There are noisy ventilators and quiet ones, and in my experience the authorities who provide ventilators for home use tend to disregard this problem.

A more deep-seated concern is whether the ventilator might break down, and what happens if it does. Upon first transition to a ventilator, this is a major worry, at least in my experience. It is still a concern for me. After all, one has been instructed that one's very life is dependent upon this machine by the side of the bed. The worry is more in the mind than in actuality; I have been using various ventilators for five years, and only once has a machine broken down. On this particular occasion it seemed Fate was smiling on me, because as I waited for 9 o'clock so I could call the company, I received a call from the engineer to tell me he was coming that very day to service my machine.

Other forms of breakdown are perhaps more worrying. In December 1990, a particularly vicious snowstorm took down almost all the power lines in the Midlands. I first knew of this when my ventilator transferred from mains electricity to its built-in dry cell battery, and gave me a little warning beep as it did so. This was at 6.50 a.m. on a

Saturday. Power was restored at 4.15 p.m. the following Thursday, a total of five nights without electricity. We have three cars (mine, my wife's, and my son's) but only one of them had a battery that was sufficiently charged to drive my ventilator. It did so for only about six hours, but a local garage came to my rescue by collecting the battery, taking it away somewhere where there was power, and charging it up. By the fifth day, the snow had gone but it was all getting rather tedious. I had acquired a petrol generator from work, but that made a colossal racket, and would only last a few hours anyway. Fortunately the power was then restored.

A concern that takes a while to diminish is the wearing of a nasal mask at night-time. Anything that is held in contact with the skin is bound to cause irritation from time to time, and one has to become accustomed to it. Initially I developed severe abrasions over the bridge of my nose, but this was resolved with a foam block to relieve the pressure on this area. The worst difficulty is when one gets a head cold. With a blocked nose, nasal ventilation becomes rather difficult! All one can do is use proprietary, or prescribed, decongestant treatments and hope that the cold does not last too long. My wife has become very tolerant of the noises, snorts, and squeaks that accompany ventilation. She sometimes wakes me up to tell me that I am snoring, which is a physical impossibility on a nasal ventilator; what she hears is air escaping from around the mask, necessitating a quick yank on the straps to tighten it up. Of course, wearing a mask has other, less obvious, implications. Being fitted with a nasal mask and yards of tubing is a most effective passion killer! It only takes a few seconds to switch it all off, however, so there is no reason for anyone faced with having to be ventilated to worry about the effects on human relationships!

TRAVELLING WITH A VENTILATOR

My objective has always been to live as normal a life as possible, and normality involves travelling. During the last five years I have been on holiday in Portugal, and am writing this in St Lucia. I have attended conferences in Germany and Holland and made two research trips to California. I love travelling, and am writing this merely to prove to anyone who is facing the prospect of being ventilated that it need not deprive you of holidays or business travel; all you need to do is plan ahead.

I admit I am more fortunate than many others, as my difficulties do not markedly restrict my mobility. On the other hand, I cannot carry anything that is even moderately heavy and I am unable to lift anything above shoulder level. That my disabilities are not obvious, on the other hand, means that people rarely offer me help. On the whole, I have found airlines very helpful. I like to deal directly with the airline rather than through a travel agent, and you have to be prepared to push hard to get a seat with maximum leg room for stowing the ventilator. I have never taken the risk of allowing the ventilator to go in the baggage hold.

CONCLUDING REMARKS

Being ventilated has given me my life back in full measure. When I was in intensive care, William Kinnear told my wife that I would one day look back and not be able to remember ever feeling better. That is most certainly true. When you are told that your future existence requires you to use a ventilator every night, the shock and foreboding are immense. If this has happened to you, just stick with it. My experience shows that life with a ventilator can be full, rewarding, and productive.

Appendix 1: Bibliography

Moxham, J. (1991). *Assisted ventilation*. BMJ Books, London.

Shneerson, J. M. (1988). *Disorders of ventilation*. Blackwell Scientific Publications, Oxford.

Gilmartin, M. E. and Make B. J. (Ed.) (1988). Mechanical ventilation in the home: issues for health care providers. *Problems in respiratory care*, vol. 1, No. 2, J. B. Lippincott, Philadelphia.

Appendix 2: Suppliers of specialist equipment

POSITIVE PRESSURE VENTILATORS AND ACCESSORIES

Lifecare, 655 Aspen Ridge Drive, Lafayette, CO 80026-9341, USA. Telephone 303/666–9234.
Lifecare Europe GmbH, Postfach 20, Hauptstrasse 60, D/8031 Seefeld 2, Germany. Telephone 49 815270728.
PLV 100 and PLV 102 positive pressure ventilators

PneuPac Ltd, Crescent Road, Luton, Bedfordshire, England, LU2 0AH. Telephone 0582 453303.
BromptonPac n-IPPV ventilator

Taema, 6 Rue Georges Besse, CE 80 92182, Antony Cedex, France. Telephone 1 40 966600.
UK agent: Deva Medical Electronics, 8 Jensen Court, Astmoor Industrial Estate, Runcorn, Cheshire, England, WA7 1PF. Telephone 0928 565836.
Monnal D and DCC positive pressure ventilators

Thomas Respiratory Systems, 33 Half Moon Lane, Herne Hill, London, England, SE24 9JX. Telephone 071 737 5881.
Nippy n-IPPV ventilator

Puritan Bennet, Portable Ventilator Division, 4865 Sterling Drive, Boulder, CO 80301, USA. Telephone 303 443 3350.
UK agent: Puritan Bennet (UK) Ltd, Unit 1, Heathrow Causeway Estate, 152–176 Great West Road, Hounslow, Middlesex, England, TW4 6JS. Telephone 081 577 1870.
Companion 2801 positive pressure ventilator, nasal pillows

Respironics Inc., 1001 Murry Ridge Drive, Murrysville, Pennsylvania 15668–8550, USA. Telephone 800 345 6443.
UK agent: Medicaid Ltd, Hook Lane, Pagham, Sussex, England, PO21 3PP. Telephone 0243 267321.
BiPap ventilator

NEGATIVE PRESSURE VENTILATOR PUMPS

DHB Tools, Althorpe Street, Leamington Spa, Warwickshire, England, CV31 2AU. Telephone 0926 426885.
Pump for paediatric negative pressure chamber

J. H. Emerson Company, 22 Cottage Park Avenue, Cambridge, MA 02140, USA. Telephone 617 864 1414.
Emerson negative pressure chest respirator

Lifecare, 655 Aspen Ridge Drive, Lafayette, CO 80026-9341, USA. Telephone 303/666–9234.
Lifecare Europe GmbH, Postfach 20, Hauptstrasse 60, D/8031 Seefeld 2, Germany. Telephone 49 815270728.
NEV 100 negative pressure ventilator

Si Plan Electronics Research, Avenue Farm Industrial Estate, Birmingham Road, Stratford-upon-Avon, England. Telephone 0789 205849.
Newmarket ENPV ventilator and Compact cuirass pump

CABINET RESPIRATORS

DHB Tools, Althorpe Street, Leamington Spa, Warwickshire, England, CV31 2AU. Telephone 0926 426885.
Adult cabinet respirator

Horner and Wells Ltd, 33 Robjohns Road, Wilford Industrial Estate, Chelmsford, Essex, England, CM1 3AN. Telephone 0245 256131.
Paediatric negative pressure chamber

PortaLung Inc., 401 East 80th Avenue, Denver, CO 80229, USA. Telephone 303 288 7575.
PortaLung cabinet respirator

JACKET RESPIRATORS

J.H. Emerson Company, 22 Cottage Park Avenue, Cambridge, MA 02140, USA. Telephone 617 864 1414.
Poncho-Wrap jacket respirator

Lifecare, 655 Aspen Ridge Drive, Lafayette, CO 80026-9341, USA. Telephone 303666–9234.

Lifecare Europe GmbH, Postfach 20, Hauptstrasse 60, D/8031 Seefeld 2, Germany. Telephone 49/815270728.

Pulmo Wrap, Nu Mo Suit, Nu Mo Jacket, Nu Mo Poncho and Nu Mo Bag jacket respirators

Orthopaedic Systems, 22–23 Oldgate, St Michael's Industrial Estate, Widnes, Cheshire, England, WA8 8TL. Telephone 051 420 3250.

X-lite for making jacket respirator shells

Peter Storm, 14 High Pavement, Nottingham, England, NG1 1HP. Telephone 0602 506911.

Airtight garments (catalogue number 104) for making jacket respirators

CUIRASS RESPIRATORS

Lifecare, 655 Aspen Ridge Drive, Lafayette, CO 80026-9341, USA. Telephone 303666–9234.

Lifecare Europe GmbH, Postfach 20, Hauptstrasse 60, D/8031 Seefeld 2, Germany. Telephone 49/815270728.

Cuirass shells

Promedics, Clarendon Road, Blackburn, England, BB1 9TA. Telephone 0254 663038.

Neoprene for cuirass belts and jacket respirator connector seals

Qbitus Products, Ryburn Mill, Hanson Lane, Halifax, England, HX1 4SD. Telephone 0422 381188.

PudgeeFoam for making cuirass shell seals

Smith and Nephew Medical Ltd, PO Box 81, Hessle Road, Hull, England, HU3 2BN. Telephone 0482 25181.

San Splint 'Synergy' for making cuirass shells, Velcro for cuirass belts

PHRENIC NERVE PACEMAKERS

Avery Laboratories Inc, 145 Rome Street, Farmingdale, New York 11735, USA. Telephone 516 293 3630.

PNEUMOBELTS

Lifecare, 655 Aspen Ridge Drive, Lafayette, CO 80026-9341, USA. Telephone 303666–9234.
Lifecare Europe GmbH, Postfach 20, Hauptstrasse 60, D/8031 Seefeld 2, Germany. Telephone · 49/815270728.
Exsufflation belt

ROCKING BEDS

RSP Manufacturing, Pegasus Way, Bowerhill Trading Estate, Melksham, Wiltshire, England, SN12 6TR. Telephone 0225 707646.

Index